BUSINESS
STATISTICS

About the Author

Brent E. Zepke was awarded a B.A. in mathematics from the University of North Carolina, an M.S. in management from Clemson University, a J.D. from the University of Tennessee, and an LL.M. from Temple University.

He has taught applied mathematics at Central Piedmont Community College; industrial engineering and statistical quality control at Greenville Technical Institute; statistics, management, business policy, business law, and management science at the University of Tennessee; and management at Temple University.

In addition, he has worked as a principal in a management consulting firm and has written many articles for academic and industrial journals.

BUSINESS STATISTICS

An Introduction

Brent E. Zepke

BARNES & NOBLE BOOKS

A DIVISION OF HARPER & ROW, PUBLISHERS

New York, Cambridge, Hagerstown, Philadelphia,
San Francisco, London, Mexico City, São Paulo, Sydney

To my wife Anne, my sons
Chad Eric and Grant Austin,
and my daughter Hollie Anne

BUSINESS STATISTICS. Copyright © 1980 by Brent E. Zepke. All rights reserved. Printed in the United States of America. No part of this book may be used or reproduced in any manner whatsoever without written permission except in the case of brief quotations embodied in critical articles and reviews. For information address Harper & Row, Publishers, Inc., 10 East 53rd Street, New York, N.Y. 10022. Published simultaneously in Canada by Fitzhenry & Whiteside Limited, Toronto.

FIRST EDITION

Designed by Eve Kirch

Library of Congress Cataloging in Publication Data

Zepke, Brent E
  Business statistics.

  Includes index.
  1. Commercial statistics. 2. Statistics.
I. Title.
HF1017.Z46 1979   519.5   78-21439
ISBN 0-06-460180-3

80 81 82 83 84 10 9 8 7 6 5 4 3 2 1

Contents

PREFACE

The objective of this book is to present business statistics in the most concise and understandable manner possible, so that one who finishes the text will be able to use effectively statistical techniques to solve business problems. The method of presentation enables the book to be used to summarize a course within a college of business administration or to supplement a text for such a course.

The book can also be effectively used by the student or business person who simply wishes to know more about statistics. No previous knowledge of statistics or higher mathematics is required, as every step of the examples is explained in great detail.

Considering the increasingly complex world of business, in which the availability of data progresses geometrically with the expanding use of electronic communications and analysis, it is apparent that the successful person must be able to organize and analyze large quantities of information. Statistics is one of the most valuable, and indeed, necessary, tools to provide an objective view of what has transpired and what may transpire in the future.

Within this book, the topics are in a sequence to provide a logical progression through the subject matter. Any tables of values that are necessary to work the problems in a particular chapter are presented within the chapter or in the Appendixes.

Because one must learn not only mathematical techniques, but also when and how to apply the techniques and how to interpret the results, many examples are included in this book. At the beginning of each chapter the topics and formulas are discussed and illustrated by numerous and diverse examples which have all been "student tested." In addition, where relevant, examples are in such a sequence as to demonstrate relationships among variables by having one or more of the variables change while the others are held constant. In each chapter there is a section entitled "Problems," which provides more example problems.

1

Introduction:
The Organization of Data

There is a logical sequence of activities that every successful person in business performs when dealing with a problem. This process can be best represented by building a model that depicts the most important aspects of a situation. Since no model can represent every aspect, each step requires assumptions about what is and what is not important. In this chapter, the general approach to problem solving by building a model is considered. Also presented here are techniques to gather, count, and organize data.

MODELS

Probability and statistics are invaluable tools in the analysis and solution of many problems that arise in business. The procedure that is generally followed in solving problems, shown in Figure 1.1 has the following steps.

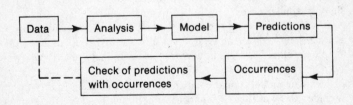

Figure 1.1

1. *Data*. Data might be obtained from measurements or from experiments, and theories may be used to generate a set of projected data. Once the data are collected, they must be counted and organized.
2. *Analysis*. The analysis of the data should be as objective as possible. Often the analysis consists of performing calculations on the data, such as to obtain averages or dispersions.
3. *Model*. The model can be an oral, written, and/or mathematical expression of the results of the analysis based on assumptions. Examples of models are normal distributions, and binomial or Poisson distributions.
4. *Predictions*. The model is used to make predictions of the results by conducting an experiment. Probability and statistical inferences may be used to make predictions.
5. *Occurrences*. Occurrences are the results of performing the experiment suggested by the model. Analysis of variance is one method for obtaining occurrences.
6. *Check of predictions with occurrences*. The final step in solving a problem must be a comparison of the predictions with the actual occurrences, by a method such as coefficient of correlation or χ^2.

The dotted line connecting the check step with the data step in Figure 1.1 indicates that the conclusions derived from the check step may be used as input to the data step to reconsider the situation if the results from the first analysis are not sufficiently accurate, if additional information is desired, or if the initial situation has changed.

To illustrate the process, we consider the results of flipping a coin 50 times.

Example 1.1

The problem is to flip a coin 50 times and then to analyze the results. To do this we apply the general problem solving format.

Let us follow each of the six solution steps through, one by one.

1. *Data*. The data are obtained from measurements (the coin is equally heavy on both faces) made from flipping the coin or other people flipping similar coins.
2. *Analysis*. We analyze the data to look for trends and similarities among them. For example, if you are using the results of a person's flipping a coin 50 times, how many heads and how many tails were obtained?
3. *Model*. The model is mathematical, verbal, or written. In this case, it includes the fact that the chances of getting a head are the same as getting a tail. This could be expressed as: Probability of a head = Probability of a tail = 0.50.

4. *Predictions.* Our prediction is that the number of heads will equal the number of tails, or that in 50 flips there will be 25 heads and 25 tails. Algebraically this is: Number of heads = Number of tails = 25.

5. *Occurrences.* We flip a coin 50 times. Note that if we wanted to be precise, we would insist that the occurrences match the assumptions made when going from the analysis to the model, which might include how the coin was held prior to flipping, the height from which it was flipped, etc.

6. *Check of predictions with occurrences.* Did 25 heads and 25 tails result, as was predicted? If not, how close to 25 of each did the experiment produce? Can this difference be allocated to chance, or was the model wrong, in which case does the model need to be revised? How close is "close enough" is a question that statistics can provide some assistance in answering.

This method for problem solving gives an overview of the study of probability and statistics. If we keep it in mind throughout the presentation of the mathematics, we will have an overview of how and when to use each topic and also how to interpret the results.

In a study of the various branches of statistics, a logical topic to commence with is techniques used to count the number of possible outcomes of an experiment.

COUNTING

Sometimes we need to know the number of possible outcomes in a given situation, for example, when the number of outcomes considered is divided by the total number of outcomes possible to determine the relative frequency or probability of an event. There are several problems of this type that we will consider individually. All such problems start with n items, from which a selection of some number of the items may be made.

For a particular situation, we must know if we are concerned with merely the total number of outcomes possible, without any regard to the order in which they occur—called a combination—or whether the order in which they occur is significant—called a permutation. For example, suppose that of 2 items A and B we select A and then B on Monday and B and then A on Tuesday. If the order in which the items were selected did not matter, then there has been one combination selected, (A and B), but if ordering does matter, then the two selection procedures were different, (A, B) and (B, A).

PERMUTATIONS: A permutation problem arises if all the items involved are to be selected and the order in which they are selected is important, in which case it is said that "the ordering matters." The first item can be selected in n ways, the second in $n - 1$ ways, and so forth.

The total number of ways in which n items may be selected is found by multiplying n by each of the positive integers (whole numbers) between itself and 0. This operation is called factorial and is indicated by n exclamation point, or $n!$.

$$n \text{ items selected in order} = n \cdot (n - 1) \cdot (n - 2) \cdots = n!$$

The result of this operation is always a positive whole number. Zero factorial ($0!$) is defined to be 1.

A permutation can also exist when we start with n items but instead of selecting all, we only select x of them, where x is less than or equal to n. For a permutation problem we find a number of arrangments, number of orderings, or number of permutations. There are n choices for the first item selected, $n - 1$ for the second item selected, and so forth.

To calculate the number of permutations of n items, selected x at a time, first calculate the number of permuations as if all n items were selected ($n!$) and then divide by the number of permutations of the items not to be selected $[(n - x)!]$.

total possible items to select $= n$
x items to be selected $= n, n - 1, n - 2, \ldots$
$n - x$ items not selected $= n - (x - 1), n - x, \ldots, 3, 2, 1$

We see that $n!$ equals the permutations from total possible items to be selected; $(n - x)!$ equals the permutations from those items not selected;

$$\text{permutations of } n \text{ items selected } x \text{ at a time} = (n)_x = \frac{n!}{(n - x)!}$$

Note that the arithmetic inside the parentheses is always performed before calculating operations outside the parentheses.

COMBINATIONS: A combination exists when we start with n items and select x of them, in any order. Because ordering does not matter, with the same number of n original items, with x items selected, there will always be fewer combinations than permutations. Some words that describe combinations are the following: number of combinations, number of groups, number of sets, arrangements do not matter.

The number of combinations is found by calculating the number of permutations and dividing by the number of excess terms that were generated for permutations but are not needed for combinations. The excess terms were generated by counting each arrangement of the same terms as a unit set for permutations, but all possible arrangements of the same terms counts as 1 in combinations.

The number of combinations of n items selected x at a time

$$= \binom{n}{x} = \frac{n!}{(n - x)! \, x!}$$

Note that the number of permutations is divided by $x!$, the number of arrangements of x items taken x at a time, to give the number of combinations because the ordering does not matter. The resulting formula may be applied directly. Also, if we selected all the items, there would be only 1 combination because there can be only 1 group if all the items are selected. In this case, x would equal n and we would have

$$\binom{x}{x} = \frac{x!}{(x-x)!\,x!} = \frac{x!}{0!\,x!} = \frac{x!}{x!} = 1$$

Example 1.2

(a) $7! = 7 \cdot 6 \cdot 5 \cdot 4 \cdot 3 \cdot 2 \cdot 1 = 5040$

(b) $(4!)(5!) = 4 \cdot 3 \cdot 2 \cdot 1 \cdot 5 \cdot 4 \cdot 3 \cdot 2 \cdot 1 = 2880$

(c) $(9!) + (3!) = (9 \cdot 8 \cdot 7 \cdot 6 \cdot 5 \cdot 4 \cdot 3 \cdot 2 \cdot 1) + (3 \cdot 2 \cdot 1) = 362,886$

(d) $(5!) - (2!) = (5 \cdot 4 \cdot 3 \cdot 2 \cdot 1) - (2 \cdot 1) = 120 - 2 = 118$

(e) $(7 - 4)! = 3! = 3 \cdot 2 \cdot 1 = 6$

(f) $(8 - 5)! = 3! = 3 \cdot 2 \cdot 1 = 6$

(g) $\dfrac{(9 - 4)!}{2!} = \dfrac{5!}{2!} = \dfrac{5 \cdot 4 \cdot 3 \cdot 2 \cdot 1}{2 \cdot 1} = \dfrac{120}{2} = 60$

(h) $\dfrac{(10 - 5)!}{(7 - 3)!} = \dfrac{5!}{4!} = \dfrac{5 \cdot \not{4} \cdot \not{3} \cdot \not{2} \cdot 1}{\not{4} \cdot \not{3} \cdot \not{2} \cdot 1} = \dfrac{5}{1} = 5$

(i) $(\frac{1}{2})!$—cannot do—must be a positive integer

(j) $(-7)!$—cannot do—must be a positive integer

(k) $(4.5)!$—cannot do—must be a positive integer

(l) $\dfrac{17!}{4!2!9!}$

$$= \frac{17 \cdot \overset{8}{\not{16}} \cdot 15 \cdot \overset{7}{\not{14}} \cdot 13 \cdot \not{12} \cdot 11 \cdot 10 \cdot \not{9} \cdot 8 \cdot \not{7} \cdot 6 \cdot \not{5} \cdot 4 \cdot \not{3} \cdot 2 \cdot 1}{\not{4} \cdot \not{3} \cdot \not{2} \cdot 1 \cdot \not{2} \cdot \not{9} \cdot \not{8} \cdot 7 \cdot 6 \cdot \not{5} \cdot \not{4} \cdot \not{3} \cdot \not{2} \cdot \not{1}}$$

$$= \frac{17 \cdot 8 \cdot 15 \cdot 7 \cdot 13 \cdot 11 \cdot 10}{1} = 20,420,400$$

Example 1.3

There are 23 employees in the machine shop. How many different groups of size 7 can be selected to be supervisors?

First, we realize that ordering does not matter and this is a combination problem, with $n = 23$ and $x = 7$.

$$\binom{n}{x} = \frac{n!}{(n-x)!\,x!} = \frac{23!}{(23-7)!7!} = \frac{23!}{16!7!}$$

$$= \frac{23 \cdot \not{22} \cdot \not{21} \cdot \not{20} \cdot 19 \cdot \not{18} \cdot 17 \cdot \not{16} \cdot \not{15} \cdot \not{14} \cdot \not{13} \cdot \not{12} \cdot 11 \cdot \not{10} \cdot \not{9}}{16 \cdot \not{15} \cdot \not{14} \cdot \not{13} \cdot \not{12} \cdot \not{11} \cdot \not{10} \cdot \not{9} \cdot 8 \cdot \not{7} \cdot 6 \cdot \not{5} \cdot \not{4} \cdot \not{3} \cdot \not{2} \cdot \not{1}}$$

$$\times \frac{8 \cdot 7 \cdot 6 \cdot 5 \cdot 4 \cdot 3 \cdot 2 \cdot 1}{7 \cdot 6 \cdot 5 \cdot 4 \cdot 3 \cdot 2 \cdot 1}$$

$$= \frac{23 \cdot 11 \cdot 19 \cdot 3 \cdot 17}{1} = 245{,}157$$

A briefer way to indicate the same arithmetic is

$$\frac{23!}{16!7!} = \frac{23 \cdot 22 \cdot 21 \cdot 20 \cdot 19 \cdot 18 \cdot 17 \cdot 16!}{7 \cdot 6 \cdot 5 \cdot 4 \cdot 3 \cdot 2 \cdot 1 \cdot 16!}$$

$$= \frac{23 \cdot 22 \cdot 21 \cdot 20 \cdot 19 \cdot 18 \cdot 17}{7 \cdot 6 \cdot 5 \cdot 4 \cdot 3 \cdot 2 \cdot 1}$$

$$= \frac{23 \cdot 11 \cdot 19 \cdot 3 \cdot 17}{1} = 245{,}157$$

The method followed for the briefer way is to write the expression term by term in the numerator until the largest expression in the denominator is reached and then to stop writing out the numbers so that this last term exactly cancels the largest term in the denominator. The other term or terms in the denominator are written out.

Example 1.4

There are 13 employees in a machine shop. How many different groups of 4 people, each with a different task, can be selected if the order of selection determines the tasks; i.e., the ordering is important. The same set of people selected in a different order would count as another group.

Since ordering matters, it is a permutation problem, with $n = 13$ and $x = 4$.

$$(n)_x = \frac{n!}{(n-x)!} = \frac{13!}{(13-4)!} = \frac{13!}{9!} = \frac{13 \cdot 12 \cdot 11 \cdot 10 \cdot 9!}{9!} = 17{,}160$$

Example 1.5

Stacy, an artist who makes mosaics, opens a box of assorted stones. There are 10 different stones, each a different type.

(a) In how many ways can she arrange the 10 stones? Here, of course, ordering matters.

Since ordering matters, it is a permutation problem with $n = 10$ and $x = 10$.

$$(n)_n = \frac{n!}{(n-x)!} = \frac{10!}{0!} = 10 \cdot 9 \cdot 8 \cdot 7 \cdot 6 \cdot 5 \cdot 4 \cdot 3 \cdot 2 \cdot 1 = 3{,}628{,}800$$

(b) How many collections of all the stones, not arranged, could she make? Here ordering would not matter.

Since ordering does not matter, it is a combination problem, with $n = 10$ and $x = 10$.

$$\binom{n}{x} = \frac{n!}{(n-x)!\,x!} = \frac{10!}{(10-10)!10!} = 1$$

(c) If ordering matters, in how many ways can she arrange 6 out of the 10 items?

Since ordering matters, it is a permutation problem, with $n = 10$ and $x = 6$.

$$(n)_x = \frac{n!}{(n-x)!} = \frac{10!}{(10-6)!} = \frac{10!}{4!}$$

$$= \frac{10 \cdot 9 \cdot 8 \cdot 7 \cdot 6 \cdot 5 \cdot 4!}{4!} = 151{,}200$$

(d) How many collections of 6 stones, not arranged, so ordering would not matter, could she make out of the box of 10 stones?

Since ordering does not matter, it is a combination problem, with $n = 10$ and $x = 6$.

$$\binom{n}{x} = \frac{n!}{(n-x)!\,x!} = \frac{10!}{(10-6)!6!} = \frac{10!}{4!6!}$$

$$= \frac{10 \cdot 9 \cdot 8 \cdot 7 \cdot 6!}{4 \cdot 3 \cdot 2 \cdot 1 \cdot 6!} = \frac{10 \cdot 3 \cdot 7}{1} = 210$$

Example 1.6

Mark receives 6 pieces of mail. In how many orders might he open them?

Since ordering matters, it is a permutation problem, with $n = 6$, $x = 6$.

$$(n)_n = n! = 6 \cdot 5 \cdot 4 \cdot 3 \cdot 2 \cdot 1 = 720$$

or

$$\frac{n!}{(n-x)!} = \frac{6!}{(6-6)!} = \frac{6!}{0!} = \frac{6!}{1} = 720$$

GROUPING DATA AND HISTOGRAMS

Often it is desirable to group data, either at the measuring stage or later, in order to deal with them more effectively. For instance, the U.S.

Census classifies data for rent paid per household in groupings such as
$0–25/mo, $26–50/mo, etc. The questionnaire used to generate the
data may ask for a specific amount of rent paid or for a check in the
appropriate square representing a specific range (say, $0–25/mo). In
either case, the statistician is able to work with data that are grouped. To
divide the data into groups it is necessary to take the total *range* (i.e., the
highest score minus the lowest score) into which the data might fall and
divide it into increments of the same size. These increments can be large
or small, depending on the accuracy desired by the statistician. Large
increments make the calculations simpler but yield less specific results. If
greater precision is desired, the increments must be made smaller. For
example, in the rent example mentioned above, increments of $100/mo
would provide for less differentiation into subgroups and hence simpler
calculations, but the results would also be more general than would those
with smaller increments of, say, $25/mo.

Once the data are grouped, the midpoint of each increment is selected
to represent that increment. The quantity of data within the increment is
called the *frequency* of the increment. Thus, each increment can be
represented by a point (its midpoint) and a *frequency distribution*. If the
number of data in each increment is divided by the total number of data,
then each frequency distribution is a percentage or fraction of the total
distribution and is called a *relative frequency distribution*. Frequency
distributions consist of positive whole numbers (integers), while relative
frequency distributions consist of positive decimals or fractions between
0 and 1. Relative frequency distributions are valuable because of their
ease of conversion to probability distributions.

Frequency distributions are sometimes represented by bar charts
called *histograms*. In a histogram the areas representing the frequency
distributions are proportional to sizes of the frequency distributions.
Also, the *cumulative function* can be generated for any given distribution; it is computed at a point by summing the values associated with the
function for the points that are less than the point and the values associated with the point. In functional notation a small letter, frequently x, is
used to denote the "independent" variable that can be assigned any
value of x. The dependent variable, whose value is a function of the
independent variable, is denoted by $f(x)$. The cumulative function is
usually denoted by a capital letter.

Example 1.7

The following is a tabulation for the number of days per month that
Senator Arvey spent in Tennessee. Give a tabulation of the cumulative
distribution for each point. Here x = months and $f(x)$ = number of days
per month.

x	1	2	3	4	5	6
f(x)	3	5	4	4	2	2

Let $F(x)$ be the cumulative function.

x	1	2	3	4	5	6
f(x)	3	5	4	4	2	2
F(x)	3	8	12	16	18	20

Example 1.8

The record shows that over the last month in a study area, the following charges were recorded:

1 charge of vagrancy was leveled at 3 persons during the first week.
2 charges of vagrancy were leveled at 4 persons during the second week.
3 charges of vagrancy were leveled at 7 persons during the third week.
5 charges of vagrancy were leveled at 8 persons during the fourth week.

Let x be the number of charges of vagrancy and $f(x)$ be the number of people charged with x vagrancies. Find the frequency distribution and cumulative distribution.

(a)   List a frequency distribution for values of x.

x	1	2	3	5
f(x)	3	4	7	8

(b)   List a relative frequency distribution for values of x.

x	1	2	3	5
f(x)	3/22	4/22	7/22	8/22

(c)   List the cumulative function for x.

x	1	2	3	5
F(x)	3	7	14	22

(d)   List the cumulative function for the relative frequency.

x	1	2	3	5
$\dfrac{F(x)}{n}$	$\dfrac{3}{22}$	$\dfrac{7}{22}$	$\dfrac{14}{22}$	1

Example 1.9

A survey is conducted to determine the number of miles commuted by salespersons per day. The results are

3.6	5.23	16.3	11.3	3.8
7.8	21.0	20.1	13.7	6.7
5.4	19.2	17.4	20.0	12.9
6.04	15.6	9.6	10.2	17.7

(a) Tabulate the frequencies and cumulative function, given a scale starting at 2.5 mi, with 5-mi increments.
(b) Construct a histogram.
(c) Graph the cumulative function.

Solutions

(a)

Increments	2.5–7.49	7.5–12.49	12.5–17.49	17.5–22.49
Frequency distribution	6	4	5	5
Cumulative distribution	6	10	15	20

(b) Histogram

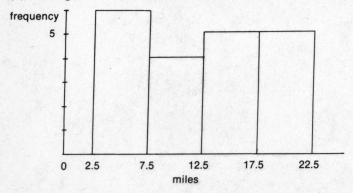

(c) Cumulative distribution

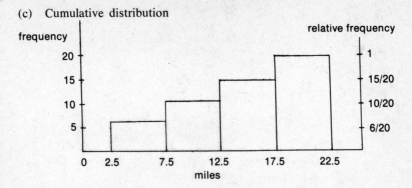

Note that in the above example the cumulative distribution and cumulative distribution of the relative frequency are shown on the same graph, with the left axis indicating one scale and the right axis indicating the other scale.

Also, note that the size of the increment selected (in this example, 5) may be decreased (to, say, 1 or 2) to increase accuracy, or may be increased (to, say, 10) to facilitate easier calculations.

Example 1.10

The Census reports the following rents paid per month. You wish to deal with 5 ranges of data, each of which must be the same size. Also draw the frequency distribution, relative frequency distribution, and cumulative frequency.

0	22	84	31	141
72	154	97	196	153
58	176	159	250	170
91	130	172	231	107
108	126	48	204	109

To find the size of the increment needed, take the difference between the highest and lowest value and divide by the number of ranges requested.

$$\text{highest} - \text{lowest} = 250 - 0 = 250$$
$$250/5 = 50$$

Increments	0–49	50–99	100–149	150–199	200–249
Frequency distribution	4	5	6	7	3
Cumulative distribution	4	9	15	22	25

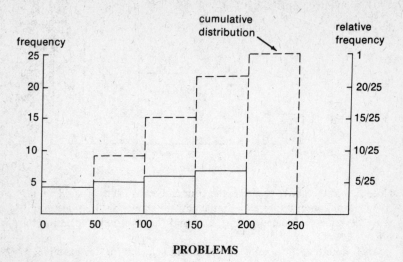

PROBLEMS

Problem 1.1

Take note of the following:

(a) $0! = 1$

(b) $4! = 4 \cdot 3 \cdot 2 \cdot 1 = 24$

(c) $6! = 6 \cdot 5 \cdot 4 \cdot 3 \cdot 2 \cdot 1 = 720$

(d) $\dfrac{5!}{3!} = \dfrac{5 \cdot 4 \cdot 3 \cdot 2 \cdot 1}{3 \cdot 2 \cdot 1} = \dfrac{120}{6} = 20$

   or $\dfrac{5 \cdot 4 \cdot 3!}{3!} = 5 \cdot 4 = 20$ by canceling the 3! from the numerator

   and denominator

(e) $\dfrac{5!}{3! \, 2!} = \dfrac{5 \cdot 4 \cdot \cancel{3!}}{2 \cdot 1 \cdot \cancel{3!}} = \dfrac{20}{2} = 10$

Problem 1.2

There are 7 books that you ask an assistant to pack, 1 at a time, and send to you. You must remove them 1 at a time from the box.

(a) If ordering matters, how many ways could you remove the books from the box?

$$n! = 7! = 5040$$

(b) If ordering does not matter, how many different groups of 5 can you select?

$$\binom{n}{x} = \frac{n!}{(n-x)!\,x!}$$

$$\binom{7}{5} = \frac{7!}{(7-5)!\,5!} = \frac{7!}{2!\,5!} = 21$$

Problem 1.3

There are 8 production lines feeding into the area where you are working. Consider only 1 product from each line.

(a) If ordering matters, in how many ways can the objects be arranged?

$$n! = 8! = 8 \cdot 7 \cdot 6 \cdot 5 \cdot 4 \cdot 3 \cdot 2 \cdot 1 = 40{,}320$$

(b) If ordering does not matter, how many groups of 6 might be selected?

$$\binom{n}{x} = \frac{n!}{(n-x)!\,x!} = \frac{8!}{(8-6)!\,6!} = \frac{8 \cdot 7 \cdot \cancel{6!}}{2!\,\cancel{6!}} = 28$$

(c) If ordering matters, in how many ways could 3 items be selected?

$$(n)_x = \frac{n!}{(n-x)!}$$

$$\frac{8!}{(8-3)!} = \frac{8 \cdot 7 \cdot 6 \cdot \cancel{5!}}{\cancel{5!}} = 336$$

Problem 1.4

A manager, E. Whitaker, wishes to pay his workers according to the "Whitaker scale," a new rate for piece rate operations that some critics say resembles industrial slavery. He hires you and instructs you to take samples of the current output and divide the output into 12 divisions. You decide that to try and moderate the Whitaker scale you will draw a histogram in the hope that the statistical presentation will illustrate the potential injustices in his initial system. The following are the data:

8	9	58	13
17	0	2	41
65	12	10	17
41	100	74	16
107	24	14	40
105	87	29	69
33	75	32	30
59	70	7	23
40	26	40	40
22	67	90	0

Note that these data are unit outputs per hour and are denoted by x.

$$\text{size of increments} = \frac{\text{highest} - \text{lowest}}{\text{number of increments}} \quad \frac{107 - 0}{12} = 9$$

The first increment ends on 8 and the next begins on 9. If the measurements had not been integers and if the first increment ended on 8, then the second increment would begin on 8.1 or 8.01 or whatever number would ensure that all the data would fall within one increment or another but not between increments. The total number of data, N, is 40.

Increments	0–8	9–17	18–26	27–35	36–44	45–53	54–62	63–71	72–80	81–89	90–98	99–107
Frequency $f(x)$	5	8	4	4	6	0	2	4	2	1	1	3
Relative frequency	$\frac{5}{40}$	$\frac{8}{40}$	$\frac{4}{40}$	$\frac{4}{40}$	$\frac{6}{40}$	0	$\frac{2}{40}$	$\frac{4}{40}$	$\frac{2}{40}$	$\frac{1}{40}$	$\frac{1}{40}$	$\frac{3}{40}$
Cumulative distribution	5	13	17	21	27	27	29	33	35	36	37	40
Cumulative distribution for relative frequency	$\frac{5}{40}$	$\frac{13}{40}$	$\frac{17}{40}$	$\frac{21}{40}$	$\frac{27}{40}$	$\frac{27}{40}$	$\frac{29}{40}$	$\frac{33}{40}$	$\frac{35}{40}$	$\frac{36}{40}$	$\frac{37}{40}$	1

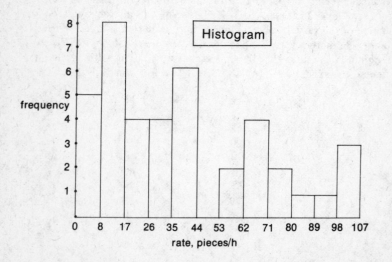

Problem 1.5

You are analyzing expenses for a corporation and arrive at the following expenses, in dollars:

50	10	5	87	12	75
47	7	8	63	26	64
32	61	12	46	37	52
41	98	41	74	55	84
19	42	17	91	66	6

Group the items into 10 groups and draw a histogram showing the frequency and relative frequency for the frequency function and the cumulative function.

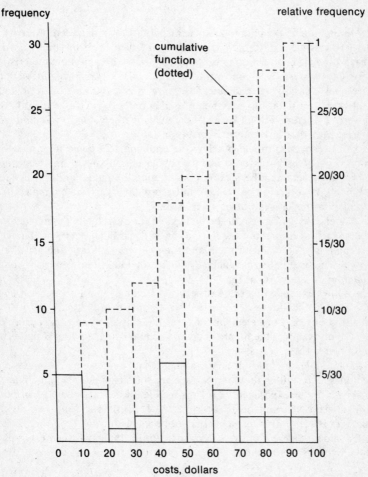

2

Probability

SETS

In organizing the data to analyze a situation, we must determine what characteristic, if any, the various items of data have in common with each other. The collections of data grouped by common characteristics are called *sets*. The concept of a set provides a shorthand way of grouping, classifying, and referring to various subgroups of the data.

Organizing data into sets helps us determine how likely it is that a given characteristic will occur, i.e., what mathematical probability of occurrence that characteristic or set has.

With respect to model building, determining sets could take place in any of the steps but is most likely to occur initially in the analysis step. Then these sets may be used in constructing the model and throughout each of the other steps. Determining probability is an example of a prediction that arises out of such models.

The elements of a set can be defined either by actually listing each of them, such as the set of (a, e, i, o, u) or by describing them by defining a characteristic common to them and them alone, such as "the set of all vowels in the English language." Each individual item, or letter, is a *member*, or *element*, of the set.

Another example of a set, say set A, is all the even numbers from 1 to 10, which could also be denoted by "set A consists of (2, 4, 6, 8, 10)." To describe one or more elements of a set the classification *subset* can be used. For example, the numbers (2, 4, 6) could be called a set B that is a subset of set A.

All possible outcomes of an experiment can be considered as a set, or *sample space*. Within the sample space are subsets of possible outcomes, or *events*. In the experiment of rolling one die, the set A, or sample space, is all the possible outcomes, or (1, 2, 3, 4, 5, 6). The subset of even numbers is (2, 4, 6). This subset could be labeled event B. If the die is rolled and an element of event B occurs, for example a 4, then it is said that "event B has occurred."

Venn diagrams are often used to illustrate sets and subsets. A Venn diagram consists of a rectangular frame, which represents the sample space, within which are unique spaces for the elements of the sample space. For example, for the experiment of the die the Venn diagram is as shown in Figure 2.1.

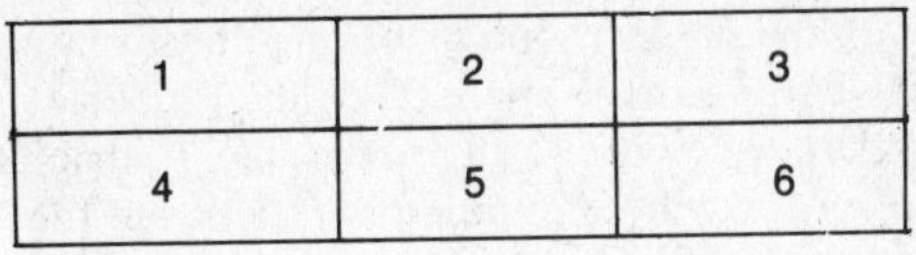

1	2	3
4	5	6

Figure 2.1

Note that in Figure 2.1 each of the possible outcomes or elements is given a unique space within the Venn diagram. (The shape of the assigned spaces is usually irrelevant.)

It is often desirable to assign a number to each outcome (event), or point of a sample space, and when this is done we may say that a function has been defined on the sample space. This function is called a *random function* or simply a *random variable*. The values assigned in this manner should be carefully chosen, as they may be useful in any desired calculations.

RELATIVE FREQUENCIES

One of the principal purposes of statistics is to transfer accurate information about a situation from one person or source to another. For example, suppose that someone told you he was successful at starting his car 7 times yesterday. While this might increase your knowledge of his efforts, it still leaves many of the circumstances unknown to you; for example, did he try 10 times or 100 times or some other number of times to start his car? The number of successes (or failures) is called the *frequency* of the event, and the number of successes (or failures) relative to the total number of occurrences is called the *relative frequency*.

$$\text{relative frequency} = \frac{\text{number of occurrences of a particular kind}}{\text{total number of occurrences}}$$

A probability of an event may be assigned by theoretical analysis. We might measure a coin, conclude that since it is balanced, the probability of a head or a tail is ½. Alternately, we might conclude that the probability is approximately equal to the relative frequency of the event over a period of time. This, unless specified otherwise, is the definition we will use. Because a particular kind of event cannot happen less than zero times or more than the total number of occurrences, the probability of an event must be between 0 and 1. Also, when the relative frequencies, or probabilities, of all the possible outcomes are added together, the sum of the numerators must equal the denominator. Hence the sum of the probabilities of all the possible outcomes must be 1. Also, if an event is defined to be the entire sample space, then the probability of that event must be 1. If they were drawn to scale, the probabilities could be calculated for each event by measuring their areas in the Venn diagram and by dividing each by the total area since the probability of an event is equal to the sum of the probabilities of the outcomes or elements that make up or are contained in the event.

DISCRETE AND CONTINUOUS PROBABILITIES

If before the experiment takes place it is possible to list all the outcomes that might occur, i.e., to chart the sample space, the experiment is considered *discrete*. In discrete problems the values that the variable can assume are specified, as are their relative frequencies. The probability function is given or obtainable. To obtain the probability of any particular value of the variable the value of the variable is plugged into the probability function, and the resulting number is the probability. If the relative frequency is given, then the probability of a particular value of the variable occurring is the relative frequency for that value of the variable. Any combination of probabilities may be obtained by plugging each individual variable value into the probability function, generating a unique probability for each point, and then combining the probabilities of the points. Thus, the probability of a particular value of the variable can be calculated and can be used directly or manipulated to achieve any desired result. The cumulative at any point is the sum of the probabilities associated with every point from the lowest value (for which the function is defined to be something besides 0) up to and including the specified point. Thus, to calculate the cumulative at any point, plug into the probability function the values of the variable one at a time starting with the lowest and continuing up to and including the specified point.

If all the possible outcomes cannot be listed before the experiment occurs, then the experiment is considered *continuous*. An experiment is either discrete or continuous. While specific outcomes cannot be defined, still the range in which the outcomes lie can usually be defined, as can the probability that an outcome lies within a given range. For example, due to the necessity of rounding off and the inaccuracies of measuring devices, it is not possible to state that a person is exactly some height, say

5.135 ft tall, but it is possible to state a range in which the height will lie, say, between 4.9 and 5.4 ft. (The normal distribution is an example of a continuous function. Determining probabilities by using this function is considered in Chapter 4.) Examples of discrete and continuous functions are shown in Figure 2.2.

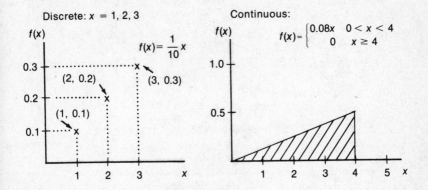

Figure 2.2

COMPLEMENTS

The set of all outcomes that are in the sample space but not in an event E is called the E *complement*, or E^c. If the probability of an event is denoted by $P(E)$, the probability of its complement is denoted by $P(E^c)$ and is equal to

$$P(E^c) = 1 - P(E)$$

A Venn diagram illustrates this relation in Figure 2.3.

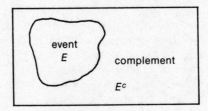

Figure 2.3

INTERSECTIONS AND UNIONS

If we specify two events, E and F, the set of elements or outcomes that are contained in both events E and F is called E *intersection* F, or $E \cap F$. The probability of E intersection F is the sum of the probabilities of all the elements common to both sets. If the sets are mutually exclusive, that is, contain no common elements, then $P(E \cap F) = 0$. Figure 2.4 shows a Venn diagram for $E \cap F$ where there are common elements indicated with cross hatching.

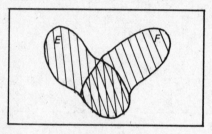

Figure 2.4

The set of elements contained in either event E or event F or in both E and F is called the *union of E and F* and is represented by $E \cap F$. There are two possible cases shown in the Venn diagram in Figure 2.5.

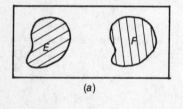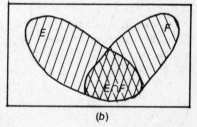

(a)

(b)

Figure 2.5

To find the probability of E union F, or $P(E \cup F)$, the probabilities of E and F should be added together. While summing these probabilities would lead to a correct solution for Figure 2.5(a), in (b), the area common to both, $P(E \cap F)$ would be counted twice when it should only be counted once, thus, the probability of this common area, or $P(E \cap F)$, must be subtracted from the sum of $P(E)$ and $P(F)$:

$$P(E \cup F) = P(E) + P(F) - P(E \cap F)$$

If the events are mutually exclusive, as in (a), and the equation still applies.

Example 2.1

Consider the possible grades received in a college course. Note that X is missed the final, I is incomplete, and W is withdrew passing. Assume that each grade is equally likely to occur.

$$\text{sample space} = (\text{set of grades possible})$$

$$= (A,\ B,\ C,\ D,\ F,\ X,\ I,\ W)$$

Let event E be making the dean's list (receiving an A or B):

$$E = (A,\ B)$$

Let event G be passing the course:

$$G = (A,\ B,\ C,\ D,\ W)$$

Assume all grades equally likely, then:

$$P\ (\text{of any grade}) = \frac{1}{\text{total number of grades}} = \frac{1}{8}$$

$$P\ (\text{of an event}) = \frac{\text{number of elements in event}}{8}$$

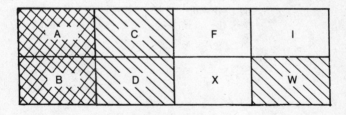

Figure 2.6

From the Venn diagram in Figure 2.6 we see that

$$P(E) = P(/\!/\!/) = P(A) + P(B) = \frac{1}{8} + \frac{1}{8} = \frac{1+1}{8} = \frac{2}{8}$$

$$P(G) = P(/\!/\!/\ ) = P(A) + P(B) + P(C) + P(D) + P(W)$$

$$= \frac{1}{8} + \frac{1}{8} + \frac{1}{8} + \frac{1}{8} + \frac{1}{8} = \frac{5}{8}$$

$$P(E^c) = P(\text{all not in } E) = P(C) + P(D) + P(F) + P(W) + P(I) + P(x)$$

$$= \frac{6}{8} = \frac{3}{4}$$

or

$$P(E^c) = 1 - P(E) = 1 - \frac{2}{8} = \frac{6}{8} = \frac{3}{4}$$

$$P(G^c) = 1 - P(G) = 1 - \frac{5}{8} = \frac{3}{8}$$

$$P(E \cap G) = P(A) + P(B) = \frac{2}{8} = \frac{1}{4}$$

$$P(E \cup G) = P(E) + P(G) - P(E \cap G) = \frac{2}{8} + \frac{5}{8} - \frac{2}{8} = \frac{5}{8}$$

CONDITIONAL PROBABILITIES

Sometimes we wish to know the probability of an event E given that some other event F has occurred. These types of problems are known as *conditional probability* problems. They are notated as $P(E/F)$. There are two types of problems that fall under this classification: dependent and independent.

DEPENDENT PROBABILITIES: Where the previous occurrence of an event F influences the probability of another event E, event E is said to be *dependent* on event F. The knowledge that event F has or has not occurred influences the probability that event E already occurred or will occur. Figure 2.7 shows the Venn diagram for this situation.

$$P(E/F) = \frac{P(E \cap F)}{P(F)}$$

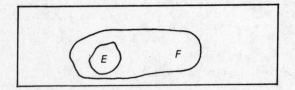

Figure 2.7

In words, the probability that E will occur given that F has already occurred is equal to the probability of E intersection F divided by the probability of F. To continue the Example 2.1 regarding grades,

$$P(E/G) \;=\; \frac{P(E \cap G)}{P(G)} = \frac{2/8}{5/8} = \frac{2}{5}$$

In other words, the probability of students making the dean's list given that they passed is 2/5. If we know that they made the dean's list and wish to know the probability that they passed, we would have:

$$P(G/E) \;=\; \frac{P(E \cap G)}{P(E)}$$

$$= \frac{2/8}{2/8} = 1$$

The interpretation of this result is that passing the course is certain, which we could know by Venn diagrams or guess by intuition.

INDEPENDENT PROBABILITIES: If the previous occurrence of an event F has no influence over the probability of event E, the two events are said to be *independent* of each other. The probability of E occurring is the same regardless of whether or not F has occurred (see Figure 2.8), or by definition when E and F are independent,

$$P(E/F) = P(E)$$

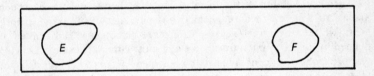

Figure 2.8

The converse is also true; that is, if $P(E/F) = P(E)$, then the events are independent. Plugging this definition into the general formula for dependent events leads to the following general formula by substituting $P(E)$ for $P(E/F)$ and solving for $P(E \cap F)$:

$$P(E/F) = \frac{P(E \cap F)}{P(F)}$$

$$P(E) = \frac{P(E \cap F)}{P(F)}$$

Therefore

$$P(E \cap F) = P(E) \cdot P(F)$$

This states that for independent events, the probability of both E and F

occurring is the probability of E multiplied by the probability of F. Again refering back to Example 2.1, we now assume that E_1 is for a student who is on the dean's list, and G_2 is for a student who is passing all his/her courses,

$$P(E_1/G_2) = P(E_1) = \frac{2}{8}$$

$$P(G_2/E_1) = P(G_2) = \frac{5}{8}$$

This indicates that their performances have no influence on each other. If the probability that they both occurred is desired, we find it as

$$P(E_1 \cap G_2) = P(E_1) \cdot P(G_2) = \left(\frac{2}{8}\right) \left(\frac{5}{8}\right) = \frac{10}{64}$$

We would intuit that the probability of both occurring together should be less than the probability of either occurring individually, which the formula substantiates.

SUMMARY

If the probability of an event is greater than 1 or is a negative number, then a mistake has been made. If by adding additional outcomes to an event the probability of the event decreases, then a mistake has been made. An outcome or an element is one particular result of an experiment, while an event is a set of one or more outcomes usually united or defined by some common trait. An event must either occur or not occur; and if it does not occur, then its complement occurs. A Venn diagram leads to the same results as the appropriate formulas.

PROBLEMS

Problem 2.1

At a baseball game, 2 hats are passed for contributions to the United Fund. When the hats reach you, they contain the following coins:

Hat 1	Hat 2
1 quarter	1 dime
1 nickel	1 nickel
1 penny	1 penny

You select one coin from each hat, so that each sample consists of a coin from hat 1 and a coin from hat 2.

(a) List the sample space.

   sample space = $[(q_1,d_2), (q_1,n_2), (q_1,p_2), (n_1,d_2), (n_1,n_2), (n_1,p_2),$
                  $(p_1,d_2), (p_1,n_2), (p_1,p_2)]$

where the letters stand for the coins for which they are initials.

(b) If all coins are drawn with equal likelihood, what is the probability of drawing 6¢?

$$P(6) = P(n_1,p_2) + P(p_1,n_2) = \frac{1}{9} + \frac{1}{9} = \frac{2}{9}$$

(c) What is the probability of drawing 30¢ or 35¢?

$$P(30 \cup 35) = P(30) + P(35)$$
$$= P(q_1,n_2) + P(q_1,d_2) = \frac{1}{9} + \frac{1}{9} = \frac{2}{9}$$

Problem 2.2

Recent MBA graduates from a leading university (UT) have supplied the following data.

Area of emphasis	Starting salary		Total
	Less than $11,000 (L)	Greater than $11,000 (G)	
Accounting (A)	800	400	1200
Economics (E)	600	450	1050
Personnel (P)	150	600	750
Total	1550	1450	3000

If 1 graduate is selected at random from this group, what is the probability that he or she is

(a) An economics major

$$P(E) = \frac{1050}{3000} = 0.35$$

(b) Not an economics major

$$P(E^c) = 1 - P(E) = 1 - 0.35 = 0.65$$

(c) An economics or personnel major

$$P(E \cup P) = P(E) + P(P) = 0.35 + \frac{750}{3000} = 0.60$$

(d) An accounting major making less than $11,000

$$P(A \cap L) = \frac{800}{3000} = 0.27$$

(e)   A UT graduate with a salary of greater than \$11,000

$$P(G \cap \text{sample space}) = \frac{1450}{3000} = 0.48$$

(f)   Given that a UT graduate makes more than \$11,000, what is the probability that he or she is a personnel major?

$$P(P/G) = \frac{P(P \cap G)}{P(G)} = \frac{600/3000}{1450/3000} = 0.41$$

Problem 2.3

Package 1 received by your purchasing department has 2 devices that are too long (L), 6 just right (R), and 2 too short (S). Package 2 has 3 too long, 4 just right, and 1 broken (B). If you draw 1 device from each package, what is the probability that you will draw the following. (Note that since the packages are separate, the drawings are independent.)

(a)   2 devices that are both too long

$$P(L_1 \cap L_2) = P(L_1) \cdot P(L_2) = \left(\frac{2}{10}\right)\left(\frac{3}{8}\right) = 0.075$$

(b)   1 too long and 1 just right

$$P(L \cap R) = P(L_1) \cdot P(R_2) + P(L_2) \cdot P(R_1) = \frac{2}{10}\left(\frac{4}{8}\right) + \frac{6}{10}\left(\frac{3}{8}\right) = 0.325$$

(c)   1 just right and 1 broken

$$P(R \cap B) = \frac{6}{10}\left(\frac{1}{8}\right) + 0 = 0.075$$

Problem 2.4

If there are 4 independent stations in a row in a production line, and the probability of an item passing through each station without becoming defective is 0.9, what is the probability of an item reaching the end nondefective?

$$P(N_1 \cap N_2 \cap N_3 \cap N_4) = P(N_1)\, P(N_2)\, P(N_3)\, P(N_4)$$
$$= (0.09)(0.9)(0.9)(0.9) = (0.9)^4 = (0.81)(0.81)$$
$$= 0.6561$$

Problem 2.5

A study was made of 300 state legislators with the following tally of results.

A = 100 who do not perceive of their role as that of a trustee

$B = 200$ who perceive of their role as that of a trustee
$C = 100$ who will consult with pressure groups
$D = 50$ who will not consult with pressure groups
$B \cap C = 20$ who perceive of their role as a trustee and will consult with pressure groups.

If 1 is to be selected from the group, give the probability of the following for that sample.

(a) $P(B) = \dfrac{200}{300} = \dfrac{2}{3}$

(b) $P(B^c) = 1 - P(B) = 1 - \dfrac{2}{3} = \dfrac{1}{3}$

(c) $P(B \cup C) = P(B) + P(C) - P(B \cap C)$

$= \dfrac{200}{300} + \dfrac{100}{300} - \dfrac{20}{300} = \dfrac{280}{300} = \dfrac{14}{15}$

(d) $P(B|C) = \dfrac{P(B \cap C)}{P(C)} = \dfrac{\dfrac{20}{300}}{\dfrac{100}{300}} = \dfrac{20}{300}\left(\dfrac{300}{100}\right) = \dfrac{1}{5}$

(e) $P(A/D) = P(A) = \dfrac{1}{3}$

(f) $P(A \cap D) = P(A/D)P(D) = \dfrac{1}{3}\left(\dfrac{1}{6}\right) = \dfrac{1}{18}$

Problem 2.6

A manager, Chad, who thinks that some of the sales personnel are "padding" their expense accounts, samples the cost of eating at a popular franchise and arrives at the following data.

Dollars spent for lunch (x)	Number of salespersons $[f(x)]$
2	8
3	18
4	30
5	10
6	14
Total	80

Determine the following probabilities:

(a) That a salesperson spent $4 for lunch.

$$P(4) = \frac{\text{number who spent \$4}}{\text{total sample size}} = \frac{30}{80} = 0.375$$

(b) That a salesperson spent $5 for lunch

$$P(5) = \frac{\text{number who spent \$5}}{\text{sample size}} = \frac{10}{80} = 0.125$$

(c) That a salesperson spent $5 or more for lunch

$$P(x \geq 5) = P(\$5) + P(\$6) = \frac{10}{80} + \frac{14}{80} = 0.125 + 0.175 = 0.300$$

(d) That a salesperson spent $4 or less for lunch

$$P(x \leq 4) = P(\$2) + P(\$3) + P(\$4) = 0.100 + 0.225 + 0.375$$
$$= 0.700$$

The result achieved by using the complement relationship is

$$P(x \leq \$4) = 1 - [P(\$5) + P(\$6)] = 1 - 0.300 = 0.700$$

(e) That a salesperson spent *less than* $4 for lunch

$$P(x < \$4) = P(\$2) + P(\$3) = 0.100 + 0.225 = 0.325$$

Note that the endpoint, in this case $4, is not included in a "less than" calculation but would be included in response to a question of "$4 or less."

(f) That a salesperson did *not* spend $3 or $4. (There are two ways to handle this.)

(1) $P(x \neq \$3) + P(x \neq \$4)$
$= P(x \neq \$3 \text{ or } X \neq \$4) = 1 - [P(\$3) + P(\$4)]$
$= 1 - (0.225 + 0.375) = 0.400$

(2) $P(x \neq \$3 \text{ or } x \neq \$4) = P(\$2) + P(\$5) + P(\$6)$
$= 0.100 + 0.125 + 0.175 = 0.400$

(g) If each time a salesperson is selected, the ticket is put back into the "pot" before another ticket is drawn, what is the probability of getting the following on successive, independent, draws?

(1) A $4 lunch followed by a $2 lunch

$P(\$2/\$4) = P(\$2) \cdot P(\$4)$ independent events

$P(\$2/\$4) = (0.100) \cdot (0.375) = 0.0375$

(2) A $5 lunch followed by a $6 lunch

$P(\$6/\$5) = P(\$6) \cdot P(\$5) = (0.175)(0.125) = 0.022$

(3) A lunch of $4 or more followed by a $2 lunch

$$P(\$2/\$4 \text{ or more}) = P(\$2) \cdot P(\$4 \text{ or more})$$
$$= (0.100)(0.625) = 0.0675$$

(4) A lunch of $2, followed by one of $5, followed by one of $6

$$P(\$6/\$5/\$2) = P(\$6) \cdot P(\$5) \cdot P(\$2)$$
$$= (0.175)(0.125)(0.100)$$
$$= 0.002$$

(h) If each time a salesperson is selected, the ticket is *not* put back into the "pot" before another ticket is drawn, what is the probability of getting the following on successive draws? (Note: these are independent events, but the sample space is reduced by 1 for the second draw to account for the ticket not put back.)

(1) A $5 lunch followed by a $2 lunch

$$P(\$2/\$5) = P(\$2) \cdot P(\$5) = \left(\frac{8}{79}\right)\left(\frac{10}{80}\right) = 0.013$$

(2) A $2 lunch followed by a $2 lunch

$$P(\$2/\$2) = P(\$2) \cdot P(\$2) = \left(\frac{8}{80}\right)\left(\frac{7}{79}\right) = 0.009$$

Problem 2.7

A study at a manufacturing plant yielded the following:

Appliance	Less than 1 defect (L)	1 or more defects (G)	Total
Radio (R)	500	200	700
TV (T)	600	400	1000
Stereo (S)	300	100	400
Total	1400	700	2100

If 1 appliance is selected, what is the probability of the selection of the following:

(a) A radio

$$P(\text{R}) = \frac{\text{number of R}}{\text{sample size}} = \frac{700}{2100} = 0.333$$

(b)   A TV

$$P(T) = \frac{\text{number of T}}{\text{sample size}} = \frac{1000}{2100} = 0.476$$

(c)   Not a stereo. (There are two methods available.)

(1) $P(S^c) = P(R) + P(T) = 0.333 + 0.476 = 0.809$

(2) $P(S^c) = 1 - P(S) = 1 - \dfrac{400}{2100} = 0.809$

(d)   A defective stereo

$$P(S \cap G) = \frac{100}{2100} = 0.0476$$

(e)   A stereo or TV

(1) $P(S \cup T) = P(S) + P(T) - P(S \cap T)$

$$= \frac{400}{2100} + \frac{1000}{2100} - 0 = 0.191 + 0.476 = 0.667$$

(2) $P(S \cup T) = 1 - P(R) = 1 - 0.333 = 0.667$

(f)   A defective radio or a nondefective TV

$P(GR \cup LT) = P(GR) + P(LT) - P(GR \cap LT)$

$$= \frac{200}{2100} + \frac{600}{2100} - 0 = 0.095 + 0.286 = 0.381$$

(g)   A defective radio, given that a radio is selected. (There are two
      solution methods available.)

(1) $P(GR) = \dfrac{\text{number of GR}}{\text{number of R}} = \dfrac{200}{700} = 0.286$

(2) $P(G/R) = \dfrac{P(G \cap R)}{P(R)} = \dfrac{200/2100}{700/2100} = \left(\dfrac{200}{2100}\right)\left(\dfrac{2100}{700}\right) = 0.286$

(h)   A nondefective TV, given that a TV is selected. (There are two
      solution methods available.)

(1) $P(LT) = \dfrac{\text{number of LT}}{\text{number of T}} = \dfrac{600}{1000} = 0.600$

(2) $P(L/T) = \dfrac{P(L \cap T)}{P(T)} = \dfrac{600/2100}{1000/2100} = \dfrac{600}{1000} = 0.600$

(i)   A nondefective appliance

$$P(L) = \frac{\text{number of L}}{\text{total appliances}} = \frac{1400}{2100} = 0.667$$

(j)   An appliance

$$P(A) = \frac{\text{number of appliances}}{\text{sample size}} = \frac{2100}{2100} = 1$$

(k)   On 2 successive draws, if the first appliance is put back before the second is selected, what is the probability (independent event) of a radio followed by a radio?

$$P(R/R) = P(R) \cdot P(R) = \left(\frac{700}{2100}\right)\left(\frac{700}{2100}\right) = 0.111$$

3

Averages and Dispersions

AVERAGES

Statistics provides ways of using single values to represent great quantities of data. For instance, one kind of value represents the central point in a range (the *median*), another represents the result that occurs most frequently(the *mode*), and a third is the familiar arithmetic average (the *mean* or *expected value*). Any or all of these quantities might be appropriate for a particular evaluation, but usually the mean is the most useful, especially when it is combined with a measure of the dispersion of the data.

To find the *median,* arrange the data in increasing order and pick the point in the center. For an odd number of items, the median is the middle item. For an even number of items, add the middle two numbers together and divide by 2.

The arithmetic average, also called the mean or expected value, is calculated by summing the items of the sample and dividing by the number of items. The results may be different from any item in the sample. For example, the mean number of children in a U.S. family is 2.3, but no family has 2.3 children.

If an item occurs more than once, a shortcut is to multiply the value of the item by the number of times it occurs and to add this number to the sum of the other items, and then to divide the sum by the total number of occurrences. The ratio of the number of times a particular item occurs to the total number of occurrences is called the *relative frequency*. If x_1, x_2, etc., represent the values whose average we are finding, then the mean value of x or the expected value of x, represented by $E(x)$, is[1]

$$E(x) = \frac{(x_1 + x_2 + x_3 + \cdots + x_N)}{N} = \sum_{i=1}^{N} \frac{x_i}{N}$$

[1] $\sum_{i=1}^{N}$ represents the sum of all the following items, starting with the first one and

Here N is the total number of occurrences and each of the items is individually represented by $x_1, x_2, x_3, \ldots, x_N$. If we group items with the same value together, then, as before, we represent the total number of items with N, but the number of different values with n. We add together each value $x_1, x_2, x_3, \ldots, x_n$ multiplied by the frequency $f(x_1), f(x_2), f(x_3), \ldots, f(x_n)$ with which it occurs and get the following expression for the mean:

$$E(x) = [x_1 f(x_1) + x_2 f(x_2) + x_3 f(x_3) + \cdots + x_n f(x_n)]/N$$

Thus,

$$E(x) = \sum_{i=1}^{n} x_i \frac{f(x_i)}{N}$$

Note that the relative frequency $\dfrac{f(x_i)}{N}$ approximates the probability of an item or event. If the data are given with the probabilities of each item or event, then the expected value can be calculated by

$$E(x) = x_1 P(x_1) + x_2 P(x_2) + x_3 P(x_3) + \cdots + x_n P(x_n)$$

$$E(x) = \sum_{i=1}^{n} x_i P(x_i)$$

The following summarizes the terms used here.

x_i = values measured	
$f(x_i)$ = frequency of each x_i	
$\dfrac{f(x_i)}{N}$ = relative frequency of x_i	
$P(x_i)$ = probability of x_i	
N = total number of occurrences	

All the above formulas yield the same results. The choice of the appropriate one should be determined by the form of the data.

---

ending with the Nth one. Note that often this is abbreviated by just Σ, which is technically not correct, but is sometimes used for convenience.

DISPERSIONS

The mean is a valuable and descriptive tool, but a more complete picture of a distribution can be obtained by using the mean with other tools. For instance, stating that two distributions have a mean of $50 might lead one to assume that the distributions are identical, when actually they might be as shown in Figure 3.1.

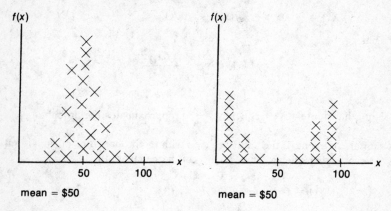

Figure 3.1

Stating that the mean is $50 adds to our knowledge of the situation, but additional information is needed for distinguishing between the two distributions. The two are different in the distances between the data points and the mean.

A measurement has been defined, called the *variance*, that expresses numerically the sum of the deviations of each of the data points from the mean. In calculating this sum the mean is subtracted from each point and that difference is squared. If this quantity were not squared, the result would be that the negative ones (where x values are more than the mean) added to the positive ones (where x values are less than the mean) could sum to zero (see Figure 3.2). After squaring, each number is multiplied by its relative frequency (which approximates its probability):

$$\text{variance} = \sum_{i=1}^{n} [x_i - E(x)]^2 \ \frac{f(x_i)}{N}$$

The variance is always a positive number whose units of measurement are the same as the mean's, only squared, so physically the units make no sense. Because the variance is the sum of each point's individual dispersion squared, the wider the spread of the distribution, the greater the numerical value of the variance. Conversely, in the unlikely case that all the points are the same, the value of the variance is 0.

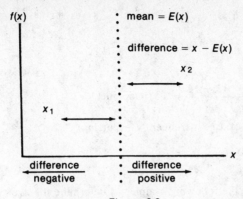

Figure 3.2

While the variance is the sum of the squared dispersions, *standard deviation* is the square root of the variance. The standard deviation is always positive and has the same units of measurement as the data and the mean. Like the variance, it also is a measure of spread or dispersion of the data about the mean: that is, the greater the spread, the larger the standard deviation; conversely, the tighter the points are packed together, the smaller the standard deviation until, if all the points concur, the standard deviation is 0.

$$\text{standard deviation} = \sqrt{\text{variance}}$$

The standard symbols used are summarized below.

Concept	Symbol
Mean	$E(x)$ or μ (mu)
Variance	σ^2 (sigma squared)
Standard deviation	σ (sigma)

Example 3.1

The data points 8, 12, 12, 12, 16, 20, 20, and 24 can be represented by the following. Determine (a) the mean, (b) the median, and (c) the mode.

x	8	12	16	20	24
$\dfrac{f(x)}{N}$	$\dfrac{1}{8}$	$\dfrac{3}{8}$	$\dfrac{1}{8}$	$\dfrac{1}{4}$	$\dfrac{1}{8}$

(a)   mean = $E(x) = \sum_{i=1}^{5} x_1 \dfrac{f(x_i)}{N} = 8\left(\dfrac{1}{8}\right) + 12\left(\dfrac{3}{8}\right) + 16\left(\dfrac{1}{8}\right)$

$$+ 20\left(\dfrac{1}{4}\right) + 24\left(\dfrac{1}{8}\right)$$

(b)   median = $\dfrac{12 + 16}{2} = 14$ (midpoint)

(c)   mode = 12 (most frequently occurring)

Example 3.2

Hollie takes the following samples to determine the average or mean time that employees spend on their breaks. What is that average time?

time spent (in minutes): 11, 12, 14, 16, 18, 19

$$E(x) = \sum_{i=1}^{n=6} x_i \frac{f(x_i)}{N} = \frac{11 + 12 + 14 + 16 + 18 + 19}{6} - 15 \text{ min}$$

Example 3.3

A sample taken to determine the average insurance carried by the workers of Rose, Inc., yielded the following (in units of $1000): 1, 5, 10, 8, 4, 8. Find (a) the mean, (b) the variance, (c) the standard deviation, and (d) the mode.

(a)   mean = $E(x) = \Sigma x \dfrac{f(x)}{N}$

$$= \frac{1}{N} \Sigma x f(x) = \frac{1}{6} [1(1) + 4(1) + 5(1) + 8(2) + 10(1)]$$

$$= \frac{36}{6} = 6$$

$6000 is the average per worker

(b)   variance = $\dfrac{1}{N} \Sigma [x - E(x)]^2 f(x)$

$$= \frac{1}{6} [(1 - 6)^2(1) + (4 - 6)^2(1) + (5 - 6)^2(1)$$

$$+ (8 - 6)^2(2) + (10 - 6)^2(1)]$$

$$= \frac{1}{6} (25 + 4 + 1 + 8 + 16) = 9$$

(c)   standard deviation $= \sqrt{\text{variance}}$
$$\sqrt{9} = 3$$
$3000 is the average insurance per worker

(d)   mode $= 8$
$8000 is the mode value of insurance per worker

Example 3.4

A survey of income for truck drivers yielded the following results in dollars per hour: 4, 2, 4, 3, 2, 4, 5, 8. What is (a) the mean of this distribution, (b) the median, (c) the mode, (d) the variance, (e) the standard deviation?

(a) $E(x) = \dfrac{1}{N} \Sigma x f(x) = \dfrac{1}{8} [(2)(2) + (3)(1) + (4)(3) + (5)(1) + (8)(1)]$

$= \left(\dfrac{1}{8}\right) 32 = 4$

mean is $4/h

(b)   median $= \$4/h$

(c)   mode $= \$4/h$

(d)   variance $= \dfrac{1}{N} \Sigma [x - E(x)]^2 f(x)$

$= \dfrac{1}{8} [(2 - 4)^2(2) + (3 - 4)^2(1) + (4 - 4)^2(3)$

$+ (5 - 4)^2(1) + (8 - 4)^2(1)]$

$= \left(\dfrac{1}{8}\right) 26 = 3.25$

(e)   standard deviation $= \sqrt{\text{variance}} = \sqrt{3.25} = \$1.80/h$

SOME THEOREMS ON EXPECTATIONS

If one adds a constant (or subtracts a constant) to each of the terms of a distribution, then the mean of the new distribution equals the mean of the original distribution plus (or minus) the constant. In algebraic terms, where c is a constant,

$$E(x + c) = E(x) + c$$

Example 3.5

Arthur Vangeli measures the time spent by his secretaries waiting for the reproduction machine to be available is (in minutes): 3, 6, 5, 12, 18, 25.

(a)   What is the average waiting time?

$$E(x) = \frac{1}{N} \Sigma \, xf(x) = \frac{1}{6} \, (3 + 6 + 5 + 12 + 18 + 25)$$

$$E(x) = 11.5 \text{ min}$$

(b)   If each secretary actually takes 3 min longer than Arthur measured, what is the average waiting time?

There are two methods available to solve this problem.

(1) add 3 to each measured time yielding for the times: 6, 9, 8, 15, 21, 28.

Then

$$E(x) = \frac{1}{N} \Sigma xf(x) = \frac{1}{6} \, (6 + 9 + 8 + 15 + 21 + 28)$$

$$E(x) = 14.5 \text{ min}$$

(2) use the equation $E(x + c) = E(x) + c$

$$E(x + 3) = 11.5 + 3 = 14.5 \text{ min}$$

Multiplication of every item in a distribution by the same constant yields a new mean that is equal to the old mean multiplied by the same constant, or,

$$c \cdot E(x) = E(cx)$$

This is true for the expected value (mean) but not true for the variance or standard deviation.

Example 3.6

Mr. Vangeli discovers that his original measurements are exactly half of the actual time.

Take the distribution from the previous example and multiply all the items by 2 and then calculate the new mean for the distribution. There are two methods available.

(1) multiply each element in the sample, to obtain 6, 12, 10, 24, 36, 50

$$E(x'') = \frac{1}{N} \sum_{i=1}^{n} x_i'' f(x_i'')$$

$$= \frac{1}{6} \sum_{i=1}^{6} x_i'' f(x_i'') = \frac{1}{6} \, (6 + 12 + 10 + 24 + 36 + 50)$$

$$E(x) = 23 \text{ min}$$

(2) by formula $E(cx) = cE(x)$ a shortcut is available, where $c = 2$ and $E(x) = 11.5$ (the expected value of the original distribution)

$$E(cx) = 2 \cdot 11.5 = 23 \text{ min}$$

Note that one of these shorter methods is available only when we have calculated a mean for a sample and then modified the sample by adding (subtracting) or multiplying (dividing) every individual measurement by a constant. This is true for the expected value (mean) but not for the variance or standard deviation.

GROUPED DATA

Once data are grouped (as discussed in Chapter 1), each increment can be represented by a point (its midpoint) and a frequency distribution (the number of data within the increment). If all the data collected are summarized and the number in each increment is divided by this total, then each frequency distribution becomes a percentage or fraction of the total distribution. This is called a *relative frequency distribution*. Hence frequency distributions consist of positive whole numbers (integers), and relative frequency distributions consist of decimals or fractions between 0 and 1. Relative frequency distributions are valuable because of their ease of conversion to probability distributions.

Once the midpoint, frequency distribution, and relative frequency distribution are known, their values can be inserted into the appropriate formulas for the mean, the variance, and the standard deviation. Since the midpoint is used to represent the increment, some loss in accuracy for computing the mean may result, but we can minimize this loss by carefully choosing the size of the increments. As we discussed in Chapter 1, a frequency distribution is sometimes represented by a histogram, and the cumulative function is generated by summing values of the data for points that are less than or equal to the points at which the cumulative function is calculated. The cumulative function is usually denoted by a capital letter.

Example 3.7

A further examination of the situation described in Problem 1.4 is presented here. A manager, E. Whitaker, wishes to pay his workers according to the "Whitaker scale," a new rate for piece rate operations that some critics say resembles industrial slavery. He hires you and instructs you to take samples of the current output and divide the output in 12 divisions. You decide that to try and moderate the Whitaker scale you will draw, as before, a histogram and also present (a) the mean, (b) the variance, (c) and the standard deviation. The data, in units of outputs per hour, are as follows:

8	9	58	13
17	0	2	41
65	12	10	17
41	100	74	16
107	24	14	40
105	87	29	69
33	75	32	30
59	70	7	23
40	26	40	40
22	67	90	0

To solve the problem, first we group the data.

$$\text{size of increments} = \frac{\text{high} - \text{low}}{\text{number of increments}} = \frac{107 - 0}{12} = 9$$

Each increment is of length 9, with the exception of the first interval, which runs from 0 to 8.

Midpoints x	4	13	22	31	40	49	58	67	76	85	94	103
Frequency $f(x)$	5	8	4	4	6	0	2	4	2	1	1	3

(a) mean $= E(x) = \sum_{i=1}^{n} \frac{x_i f(x_i)}{N} = \frac{1}{N} \sum_{i=1}^{n} x_i f(x_i)$

$$= \frac{1}{40} [(4)(5) + (13)(8) + (22)(4) + (31)(4) + (40)(6)$$

$$+ (49)(0) + (58)(2) + (67)(4) + (76)(2) + (85)(1)$$

$$+ (94)(1) + (103)(3)]$$

$$= \frac{1}{40} (1600) = 40.0 \text{ outputs/h}$$

(b) variance $= \sigma^2 = \frac{1}{N} \sum_{i=1}^{n} [x_i - E(x)]^2 f(x_i) = \frac{1}{40} \sum_{i=1}^{12} (x_i - 40)^2 f(x_i)$

$$= \frac{1}{40} [(4 - 40)^2(5) + (13 - 40)^2(8) + (22 - 40)^2(4)$$

$$+ (31 - 40)^2(4) + (40 - 40)^2(6) + (49 - 40)^2(0)$$

$$+ (58 - 40)^2(2) + (67 - 40)^2(4) + (76 - 40)^2(2)$$

$$+ (85 - 40)^2(1) + (94 - 40)^2(1) + (103 - 40)^2(3)$$

$$= \frac{1}{40} (6480 + 5832 + 1296 + 324 + 0 + 0 + 648$$

$$+ 2916 + 2592 + 2025 + 2916 + 11907)$$

$$= 923.4$$

(c) standard deviation $= \sigma = \sqrt{\sigma^2} = \sqrt{923.4} = 30.4$ outputs/h

Note that for E. Whitaker to assign pay rates at a unit rate to have the "average" person average \$5.00/h, an output of 40.0 units/h should be paid \$5.00/h. If he also wishes to establish a progressive wage structure, he could use the mean in conjunction with the standard deviation to generate such a structure.

PROBLEMS

Problem 3.1

Twenty students at the University of Tennessee had the following grade point averages:

3.4	2.0	2.5	3.0
2.1	2.5	2.4	2.8
2.7	1.9	2.3	2.6
3.0	3.2	2.2	3.8
1.8	2.6	1.8	0.8

Find (a) the mean, (b) the variance, (c) the standard deviation, and (d) the median grade.

(a) mean $= E(x) = \dfrac{1}{N} \displaystyle\sum_{i=1}^{n} x_i f(x_i)$

$$= \frac{1}{20} [(3.4)(1) + (2.1)(1) + (2.7)(1) + (3.0)(2) + (1.8)(2)$$

$$+ (2.0)(1) + (2.5)(2) + (1.9)(1) + (3.2)(1) + (2.6)(2)$$

$$+ (2.4)(1) + (2.3)(1) + (2.2)(1) + (2.8)(1) + (3.8)(1)$$

$$+ (0.8)(1)]$$

$$= \left(\frac{1}{20}\right) 49.4 = 2.47 \text{ grade points}$$

(b) variance $= \dfrac{1}{N} \displaystyle\sum_{i=1}^{n=16} [x_i - E(x)]^2 f(x_i)$

$$= \frac{1}{20} \sum_{i=1}^{n=16} (x - 2.47)^2 f(x)$$

$$= \frac{1}{20} [(3.4 - 2.47)^2(1) + (2.1 - 2.47)^2(1) + (2.7 - 2.47)^2(1)$$

$$+ (3.0 - 2.47)^2(2) + (1.8 - 2.47)^2(2) + (2.0 - 2.47)^2(1)$$

$$+ (2.5 - 2.47)^2(2) + (1.9 - 2.47)^2(1) + (3.2 - 2.47)^2(1)$$

$$+ (2.6 - 2.47)^2(2) + (2.4 - 2.47)^2(1) + (2.3 - 2.47)^2(1)$$
$$+ (2.2 - 2.47)^2(1) + (2.8 - 2.47)^2(1) + (3.8 - 2.47)^2(1)$$
$$+ (0.8 - 2.47)^2(1)]$$

$$= \frac{1}{20} (0.86 + 0.14 + 0.05 + 0.56 + 0.90 + 0.22 + 0.32$$

$$+ 0.53 + 0.03 + 0.01 + 0.03 + 0.07 + 0.11 + 1.77$$
$$+ 2.79)$$

$$= \left(\frac{1}{20}\right) 8.39 = 0.416$$

(c) standard deviation $= \sqrt{0.416} = 0.65$ grade points
(d) To find the median grade the grades must be put into increasing order. Because there is an even number of items, the middle 2 items must be averaged.
0.8, 1.8, 1.8, 1.9, 2.0, 2.1, 2.2, 2.3, 2.4, 2.5, 2.5, 2.6, 2.6, 2.7, 2.8, 3.0, 3.0, 3.2, 3.4, 3.8
The two middle numbers are both 2.5, so their average is 2.5. The median is 2.5 grade points.

Problem 3.2

For the following distribution determine (a) the mean, (b) the median, (c) the mode, (d) the variance, and (e) the standard deviation.

x	5	7	9	11	13
Frequency $f(x)$	6	4	5	7	2

The relative frequencies must first be generated. Note that there are 24 items.

x	5	7	9	11	13
$f(x)$	6	4	5	7	2
$\dfrac{f(x)}{N}$	$\dfrac{6}{24}$	$\dfrac{4}{24}$	$\dfrac{5}{24}$	$\dfrac{7}{24}$	$\dfrac{2}{24}$

(a) mean $= E(x) = \Sigma \, x \, \dfrac{f(x)}{N} = 5\left(\dfrac{6}{24}\right) + 7\left(\dfrac{4}{24}\right) + 9\left(\dfrac{5}{24}\right)$

$$+ 11\left(\dfrac{7}{24}\right) + 13\left(\dfrac{2}{24}\right)$$

$$= \frac{1}{24} [5(6) + 7(4) + 9(5) + 11(7) + 13(2)]$$

$$= \frac{1}{24} (30 + 28 + 45 + 77 + 26)$$

$$= 8.6$$

(b) To find the median, there is an even number of items, 24, and the middle 2 items both are 9, so the median is 9.

(c) mode is 11

(d) variance $= \Sigma [x - E(x)]^2 \frac{f(x)}{N} = \frac{1}{N} \Sigma [x - E(x)]^2 f(x)$

$$= \frac{1}{24} \Sigma (x - 8.6)^2 f(x)$$

$$= \frac{1}{24} [(5 - 8.6)^2 6 + (7 - 8.6)^2 4 + (9 - 8.6)^2 5$$

$$+ (11 - 8.6)^2 7 + (13 - 8.6)^2 2]$$

$$= \frac{1}{24} (77.76 + 10.24 + 0.8 + 40.32 + 38.72) = 6.99$$

(e) standard deviation $= \sqrt{\text{variance}} = \sqrt{6.99} = 2.64$

Problem 3.3

In the roll of a die, what is (a) the expected value, (b) the variance, and (c) the standard deviation?

Let x = values of the die and $P(x)$ = probability of a particular value of x.

x	1	2	3	4	5	6
$P(x)$	1/6	1/6	1/6	1/6	1/6	1/6

(a) $E(x) = \sum_{i=1}^{6} x_i P(x_i) = 1(1/6) + 2(1/6) + 3(1/6) + 4(1/6) + 5(1/6)$

$$+ 6(1/6)$$

$$= 1/6 + 2/6 + 3/6 + 4/6 + 5/6 + 6/6 = 21/6 = 3.5$$

So if a die were repeatedly rolled, we could expect to average 3.5 per roll.

(b) $\sigma^2 = \frac{1}{N} \sum_{i=1}^{6} [x_i - E(x)]^2 f(x_i) = \sum_{i=1}^{6} (x - 3.5)^2 \frac{f(x_i)}{6}$

$$= (1 - 3.5)^2 1/6 + (2 - 3.5)^2 1/6 + (3 - 3.5)^2 1/6$$

$$+ (4 - 3.5)^2 1/6 + (5 - 3.5)^2 1/6 + (6 - 3.5)^2 1/6$$

$$= (1/6)(6.25 + 2.25 + 0.25 + 0.25 + 2.25 + 6.25)$$

$$= 2.92$$

(c)   standard deviation $= \sqrt{\text{variance}} = \sqrt{2.92} = 1.71$

Problem 3.4

Two dice are rolled together. What are (a) the expected value, (b) the variance, (c) the standard deviation, (d) the median, and (e) the mode of the theoretical distribution?

x = value of the two dice $P(x)$ = probability of each x

The sample space looks like this:

x (number rolled)	2	3	4	5	6	7	8	9	10	11	12
$P(x)$	$\frac{1}{36}$	$\frac{2}{36}$	$\frac{3}{36}$	$\frac{4}{36}$	$\frac{5}{36}$	$\frac{6}{36}$	$\frac{5}{36}$	$\frac{4}{36}$	$\frac{3}{36}$	$\frac{2}{36}$	$\frac{1}{36}$

(a)   expected value $= E(x) = \Sigma x P(x)$
$E(x) = 2(1/36) + 3(2/36) + 4(3/36) + 5(4/36) + 6(5/36)$

$+ 7(6/36) + 8(5/36) + 9(4/36) + 10(3/36) + 11(2/36)$

$+ 12(1/36) = 1/36 \ (2 + 6 + 12 + 20 + 30 + 42 + 40$

$+ 36 + 30 + 22 + 12)$

$= 7$

(b)   variance $= \Sigma \ [x - E(x)^2 P(x) = \Sigma \ (x - 7)^2 P(x)$

$= (2 - 7)^2(1/36) + (3 - 7)^2(2/36) + (4 - 7)^2(3/36)$

$+ (5 - 7)^2(4/36) + (6 - 7)^2(5/36) + (7 - 7)^2(6/36)$

$+ (8 - 7)^2(5/36) + (9 - 7)^2(4/36) + (10 - 7)^2(3/36)$

$+ (11 - 7)^2(2/36) + (12 - 7)^2(1/36)$

$= 1/36 \ [25(1) + 16(2) + 9(3) + 4(4) + 1(5) + 0(6)$

$+ 1(5) + 4(4) + 9(3) + 16(2) + 25(1)]$

$= 5.83$

(c)   standard deviation $= \sqrt{\text{variance}} = \sqrt{5.83} = 2.41$
(d)   median = 7 since the control point is 7 and even though there is an even number of points, both points in the center are 7.
(e)   mode = 7

Problem 3.5

Scott studies the telephone bills for his employer and finds the following numbers of minutes were spent per phone call:

10	9	13	7	1
15	21	17	5	16
7	11	14	2	2

What are (a) the mean, (b) the variance, and (c) the standard deviation?

(a)   mean $= E(x) = \sum_{i=1}^{n} x_i \frac{f(x_i)}{N} = \frac{1}{N} \sum_{i=1}^{n} x_i f(x_i)$

$E(x) = \frac{1}{15} \sum_{i=1}^{15} x_i f(x_i) = (1/15) \, [10(1) + 15(1) + 7(2) + 9(1)$

$+ 21(1) + 11(1) + 13(1) + 17(1) + 14(1) + 5(1) + 2(2)$

$+ 1(1) + 16(1)]$

$= (1/15)150 = 10$ min/call

(b)   variance $= \frac{1}{N} \sum_{i=1}^{n} [x_i - E(x)]^2 f(x_i) = \frac{1}{15} \sum_{i=1}^{15} (x_i - 10)^2 f(x_i)$

$= (1/15) \, [(10 - 10)^2 1 + (15 - 10)^2 1 + (7 - 10)^2 2$

$+ (9 - 10)^2 1 + (21 - 10)^2 1 + (11 - 10)^2 1$

$+ (13 - 10)^2 1 + (17 - 10)^2 1 + (14 - 10)^2 1$

$+ (5 - 10)^2 1 + (2 - 10)^2 2 + (1 - 10)^2 1 + (16 - 10)^2 1]$

$= (1/15) \, (0 + 25 + 18 + 1 + 121 + 1 + 9 + 49 + 16$

$+ 25 + 128 + 81 + 36)$

$= (1/15)(510) = 34$

(c)   standard deviation $= \sigma = \sqrt{\text{variance}} = \sqrt{34} = 5.83$ min/call
If the standard deviation were 0, Scott would know that all calls took the same time, which time would be the mean, or 10 min/call. The greater the standard deviation, the greater the difference in the time spent per individual phone call.

Problem 3.6

Suppose in the previous example, Scott was mistaken: every call actually took 5 min more than he initially thought because he forgot to count the time spent dialing and waiting for the other party to answer. How would this affect the mean? The new sample space is

15	14	18	12	6
20	26	22	10	21
12	16	19	7	7

$$E(x') = \frac{1}{N} \sum_{i=1}^{n} x_i' f(x_i')$$

$$= \frac{1}{15} \sum_{i=1}^{15} x_i' f(x_i')$$

$$= (1/15) [15(1) + 20(1) + 12(2) + 14(1) + 26(1) + 16(1)$$

$$+ 18(1) + 22(1) + 19(1) + 10(1) + 7(2) + 6(1) + 21(1)$$

$$= (1/15)(225) = 15 \text{ min/call}$$

$E(x'_i)$ is 5 min larger than the old $E(x)$ of 10 min, and 5 was the amount added to each of the x terms. In this example,

$$E(x + 5) = E(x) + 5$$

or

$$E(x + 5) = 15$$

$$E(x) + 5 = 10 + 5$$

Problem 3.7

Take the distribution from the previous example and multiply all the items by 2 and then calculate the new mean for the distribution.

30	28	36	24	12
40	52	44	20	42
24	32	38	14	14

$$E(x'') = \frac{1}{N} \sum_{i=1}^{n} x_i'' f(x_i'') = \frac{1}{15} \sum_{i=1}^{15} x_i'' f(x_i'')$$

$$= (1/15) [30(1) + 40(1) + 24(2) + 28(1) + 52(1) + 32(1)$$

$$+ 36(1) + 44(1) + 38(1) + 20(1) + 14(2) + 12(1) + 42(1)]$$

$$= (1/15)(450) = 30 \text{ min/call}$$

By formula, a shortcut is available:

$$cE(x) = E(cx)$$

where $c = 2$ and $E(x) = 15$ (the expected value of the original distribution)

$$2 \cdot 15 = E(2x) = 30 \text{ min/call}$$

Note that one of these shorter methods is available only when we have calculated a mean for a sample and then the sample is modified by adding (subtracting) or multiplying (dividing) every individual measurement by a constant. This is true for the expected value (mean) but *not* for the variance or standard deviation.

Problem 3.8

In order to evaluate their sales personnel, Ben Yung Auto asks Alma to determine the indicated terms for this distribution, where x equals the number of stops per month for each salesperson:

105	130
110	120
120	135

Find (a) the mean, (b) the variance, (c) the standard deviation.

(a) mean = $(x) = \dfrac{1}{N} \sum_{i=1}^{n} x_i f(x_i) = \dfrac{1}{6} \sum_{i=1}^{6} x_i f(x_i)$

$= (1/6) [105(1) + 110(1) + 120(2) + 130(1) + 135(1)]$

$= 120$ (stops/month)/salesperson

(b) variance = $\sigma^2 = \dfrac{1}{N} \sum_{i=1}^{n} [x_i - E(x)]^2 f(x_i)$

$= (1/6) \sum_{i=1}^{6} (x_i - 120)^2 f(x_i)$

$= (1/6) [(105 - 120)^2 1 + (110 - 120)^2 1 + (120 - 120)^2 2$

$+ (130 - 120)^2 1 + (135 - 120)^2 1]$

$= (1/6)(225 + 100 + 0 + 100 + 225) = 108.33$

(c) standard deviation = $\sigma = \sqrt{\sigma^2} = \sqrt{108.33}$

$= 10.41$ stops/month/salesperson

Problem 3.9

You are analyzing expenses for a corporation and arrive at the following expense items:

50	87	10	12	5	75
47	63	7	26	8	64
32	46	61	37	12	52
41	74	98	55	41	84
19	91	42	66	17	6

Group the items into 10 groups and draw a histogram showing frequency and relative frequency for the function and the cumulative function. Also calculate (a) the mean, (b) the variance, and (c) the standard deviation from the grouped data.

Increments	0–10	11–20	21–30	31–40	41–50	51–60	61–70	71–80	81–90	91–100
Frequency $f(x)$	5	4	1	2	6	2	4	2	2	2
Relative frequency $f(x)/N$	$\frac{5}{30}$	$\frac{4}{30}$	$\frac{1}{30}$	$\frac{2}{30}$	$\frac{6}{30}$	$\frac{2}{30}$	$\frac{4}{30}$	$\frac{2}{30}$	$\frac{2}{30}$	$\frac{2}{30}$
Cumulative frequency	5	9	10	12	18	20	24	26	28	30
Cumulative relative frequency	$\frac{5}{30}$	$\frac{9}{30}$	$\frac{10}{30}$	$\frac{12}{30}$	$\frac{18}{30}$	$\frac{20}{30}$	$\frac{24}{30}$	$\frac{26}{30}$	$\frac{28}{30}$	$\frac{30}{30}$

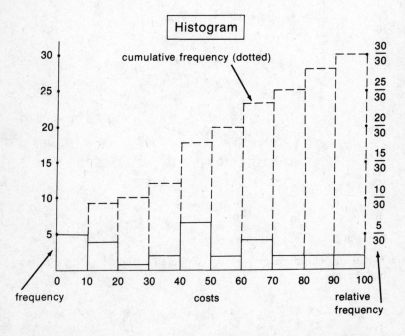

To read the histogram, the scale on the left indicates frequency and cumulative frequency and the scale on the right indicates relative frequency and cumulative relative frequency. This style of diagram is a very convenient method of representing a great deal of information in a compact manner.

For the calculations necessary, let the midpoint represent the block. For example, the 0–10 block, with a frequency of 5, has a midpoint of 5.

Midpoints of blocks	5	15	25	35	45	55	65	75	85	95
Frequencies	5	4	1	2	6	2	4	2	2	2
Relative frequencies	$\frac{5}{30}$	$\frac{4}{30}$	$\frac{1}{30}$	$\frac{2}{30}$	$\frac{6}{30}$	$\frac{2}{30}$	$\frac{4}{30}$	$\frac{2}{30}$	$\frac{2}{30}$	$\frac{2}{30}$

(a) mean $= E(x) = (1/N)\Sigma x f(x) = 5(5/30) + 15(4/30) + 25(1/30)$

$$+ 35(2/30) + 45(6/30) + 55(2/30) + 65(4/30)$$

$$+ 75(2/30) + 85(2/30) + 95(2/30)$$

$$= 1/30 \, [5(5) + 15(4) + 25(1) + 35(2) + 45(6)$$

$$+ 55(2) + 65(4) + 75(2) + 85(2) + 95(2)]$$

$$= (1/30)(1330) = 44.3$$

(b) variance $= \sigma^2 = \dfrac{1}{N} \sum_{i=1}^{n} [x - E(x)]^2 f(x) = \dfrac{1}{30} \Sigma \, (x - 44.3)^2 f(x)$

$$= (1/30) \, [(5 - 44.3)^2 5 + (15 - 44.3)^2 4 + (25 - 44.3)^2 1$$

$$+ (35 - 44.3)^2 2 + (45 - 44.3)^2 6 + (55 - 44.3)^2 2$$

$$+ (65 - 44.3)^2 4 + (75 - 44.3)^2 2 + (85 - 44.3)^2 2$$

$$+ (95 - 44.3)^2 2]$$

$$= (1/30) \, (7722 + 3434 + 372 + 173 + 3 + 229 + 1714$$

$$+ 1885 + 3313 + 5141)$$

$$= 800$$

(c) standard deviation $= \sigma = \sqrt{\sigma^2} = \sqrt{800} = 28.3$

Problem 3.10

You start a new job as personnel manager and discover on the first day that the previous manager, now in Argentina, had a system of job classification for the 36 secretaries that consisted of 25 separate levels of pay. For ease of administration you wish to consolidate the classifications into just 5 levels. Having studied statistics, you decide to obtain a graphic representation of the system by drawing a histogram and from this to calculate (a) the mean, (b) the variance, and (c) the standard deviation.

The 36 secretaries were paid according to the following rates (in dollars per half-day):

4	5	5	18	3	17
7	12	3	14	7	2
18	15	13	12	3	24
9	22	17	15	24	2
23	4	17	18	15	13
11	9	20	11	14	8

$$\text{size of increments} = \frac{\text{high} - \text{low}}{\text{number of increments}} = \frac{25 - 0}{5} = 5$$

Increments	1–5	6–10	11–15	16–20	21–25
Frequency $f(x)$	9	5	11	7	4
Relative frequency $\frac{f(x)}{N}$	$\frac{9}{36}$	$\frac{5}{36}$	$\frac{11}{36}$	$\frac{7}{36}$	$\frac{4}{36}$
Cumulative frequency	9	14	25	32	36
Cumulative relative frequency	$\frac{9}{36}$	$\frac{14}{36}$	$\frac{25}{36}$	$\frac{32}{36}$	$\frac{36}{36}$

Use the midpoints of each increment to represent the increments.

Midpoints	3	8	13	18	23
Frequency	9	5	11	7	4

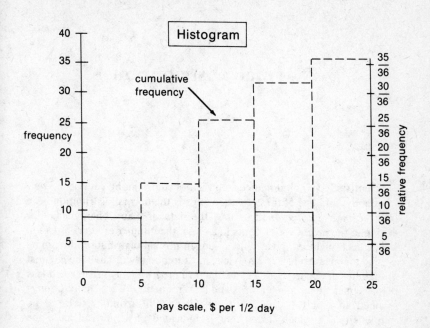

(a)   mean $= E(x) = \dfrac{1}{N} \, \Sigma \; x \, f(x) = \dfrac{1}{36} \, [3(9) + 8(5) + 13(11)$
$+ \; 18(7) + 23(4)]$
$= \$11.89/\tfrac{1}{2} \text{ day}$

(b)   variance $= \; \sigma^2 = \dfrac{1}{N} \, \Sigma \; [x - E(x)]^2 f(x) = \dfrac{1}{36} \, \Sigma \; (x - 11.9)^2 \, f(x)$

$= (1/36) \, [(3 - 11.9)^2 9 + (8 - 11.9)^2 25 + (13 - 11.9)^2 11$

$+ \; (18 - 11.9)^2 7 + (23 - 11.9)^2 4]$

$= (1/36) \, (712.9 + 76.0 + 13.3 + 260.5 + 492.8)$

$= 43.2$

(c)   standard deviation $= \sigma = \sqrt{43.2} = \$6.60/\tfrac{1}{2} \text{ day}$

4
Normal Distribution

Scientists have noted that certain relationships among data occur very frequently, and one of these relationships is the normal distribution. The term *normal distribution* describes the data of many phenomena: for example, the heights of students in a class, the number of hours a person sleeps each night in a year, and so on. In normal distribution the points around the mean value tend to occur most frequently or have the greatest probability of occurrence. Values that are progressively greater and less than the mean tend to occur less and less frequently. If the distribution is graphed so that the data points all along a horizontal axis and the frequency with which they occur is indicated on a vertical axis, the normal distribution looks like a bell:

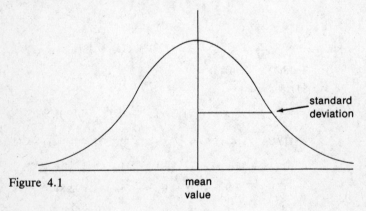

Figure 4.1   mean value

Different phenomena have different bell curves. The value of the mean determines where on a number line the center of the bell is situated, and the standard deviation determines how wide or narrow the bell is. The greater the standard deviation, the larger is the spread of data points and the wider is the bell. The standard deviation occurs on the graph as the horizontal distance between the center axis and curve at a point of inflection (where the curve changes from concave to convex).

The general form of the equation for the normal distribution is

$$f(x) = \frac{1}{\sigma\sqrt{2\pi}}\, e^{-\frac{(x-\mu)^2}{2\sigma^2}} \quad \text{for } -\infty < x < +\infty$$

where π and e are constants, μ is the mean, and σ is the standard deviation.

STANDARD NORMAL DISTRIBUTION

The area under the curve between any two points indicates the probability of the function having a value between the two points. It can be calculated by integration of the function between the two points, which is quite difficult.[1] Integration of one particular normal distribution, that of *the standard normal distribution,* has been completely calculated, and the results are set up in Table 4.1. Since using the table is much simpler than doing the integration, most problems involving the normal distribution are solved by converting the given normal distribution into the relevant points on the standard normal distribution and then using the table. By means of the following equation we can convert a given distribution to the standard normal distribution, which makes it possible for us to use the standard table:

$$z = \frac{x - \mu}{\sigma}$$

where μ is the mean, σ is the standard deviation, x is a point along the horizontal axis in the original scale, and z represents the point along the horizontal axis in the converted scale. In Figure 4.2 is shown a graph of the standard normal distribution. The percentages represent the proportions of the area under the curve between ± 1, ± 2, and ± 3 standard deviations. Thus the probability of a value falling between -1 standard deviation and $+1$ standard deviation from the mean is 0.68.

Example 4.1

Suppose someone scored 580 on the Graduate Test for Business (GBT), the mean score on the exam was 500, and the standard deviation

---

[1] Integration is studied in calculus. In Chapter 14 integration of certain elementary functions, but not the normal function, is considered.

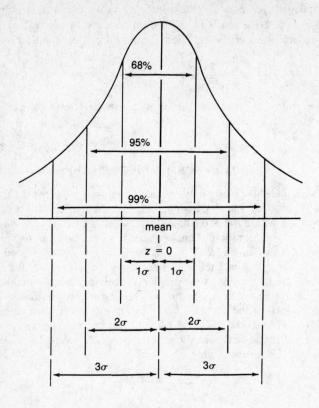

Figure 4.2

was 100. We wish to represent these data on a graph and find a variety of probabilities by means of the standard normal curve.

If we assume a bell-curve distribution, our graph of the data looks like Figure 4.3. We use the relation $z = (x - \mu)/\sigma$ to convert the data into a form with which we can use the normal standard distribution. On the z scale, the mean is 0 and x_1 becomes 0.8. Since we are dividing by the standard deviation value in old units, its new value is 1 in the new units, which we call standard deviation units or z units.

$$z_1 = \frac{x_1 - \mu}{\sigma}$$

$$= \frac{580 - 500}{100}$$

$$= 0.8$$

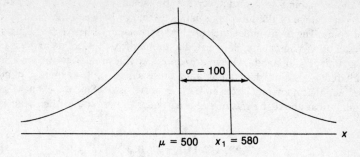

Figure 4.3

We represent these values on the standard normal curve in Figure 4.4.

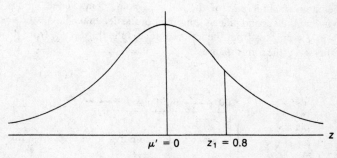

Figure 4.4

Note that the curves are the same and the points are located in the same relative positions. In fact, rather than drawing both curves, we could draw just one and provide two sets of labels (Figure 4.5), the second, the z values, in standard deviation units.

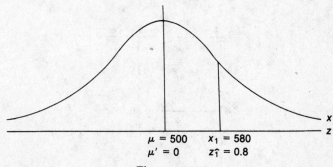

Figure 4.5

The probability of scoring between the mean and a particular point is the percentage of the total area under the curve between the mean and the point. The areas under the standard normal curve between the mean and all other points have been listed in tabular form for points expressed as z statistics, in standard deviation units. Any desired area—hence also probability—is found by manipulating these values while keeping in mind the fact that the total area under the curve is 1. Since the curve is symmetrical about the mean, the area under either side of the mean is 0.5.

USE OF THE z TABLE

Appendix I is a table that indicates the proportion of the total area under the normal curve contained in the segment between the mean and another value in standard deviation units, or z statistics. The table is developed for z statistics of three significant figures, with the place to the immediate left and right of the decimal point being listed on the left margin and the second place to the right of the decimal being listed across the top. Two places from the left margin and the third place from the top together give the three figures.

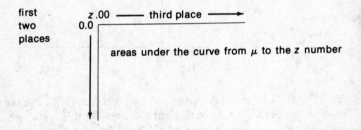

For example, let us look for $z = 0.86$.

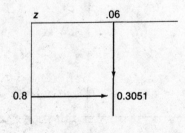

Where the row crosses the column, the number is the area under the curve between μ and $z = 0.86$.

For x values less than the mean, the z statistic is negative. If it is sought properly, the negative value results automatically. The negative sign is ignored, however, when the values are looked up in the table; that is, the area associated with a z number of $+1.78$ is the same as that associated with -1.78. The negative sign is accounted for when we decide whether to add or subtract the area from another area or 0.5.

Example 4.2

The results of the Graduate Test for Business (GTB) indicate a mean of 500 and a standard deviation of 100. What is the probability of a given person's test score (x) being one of the following.

(a) Between 500 and 550, that is, $500 \leq x \leq 550$

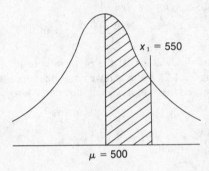

Figure 4.6

The probability of x being between 500 and 550 is equal to the area under the normal curve between 500 and 550, as shaded in Figure 4.6. The points in question must be transposed to z numbers so the table for the normal distribution can be used to provide the area under the curve between μ and x.

$P(500 \leq x \leq 550)$ = area under curve between 500 and 550

Converting the points to z numbers, we get

$$z = \frac{x - \mu}{\sigma} = \frac{550-500}{100} = \frac{50}{100} = 0.50$$

From Appendix I if we represent the area from the mean to $z = 0.5$ by $A_{z=0.5}$, we have

$$A_{z=0.5} = 0.1915$$

Thus, $P(500 < x < 550) = 0.1915$

(b) Between 450 and 500

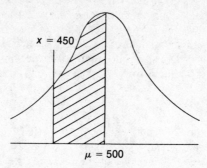

Figure 4.7

Converting to z numbers yields

$$z = \frac{x - \mu}{\sigma} = \frac{450 - 500}{100} = -0.50$$

The minus sign indicates that the point in question is to the left of the mean (Figure 4.7), but when we use Appendix I to find the areas under the curve, the minus sign is not considered; that is,

$$A_{z=0.5} = A_{z=-0.5} = 0.1915$$

Thus, $P(450 < x < 500) = 0.1915$

(c) Between 550 and 600

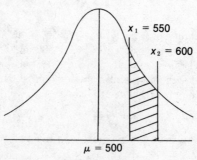

Figure 4.8

Note that here one boundary for the desired area is not the mean (Figure 4.8). Appendix I gives areas from the mean to a particular value, so the answer must be pieced together from the information obtainable from the table.

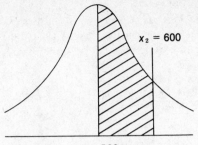

Figure 4.9 $\mu = 500$

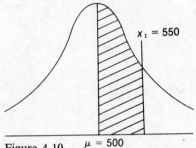

Figure 4.10 $\mu = 500$

Our technique is to find the area from the mean to $x = 600$ (represented in Figure 4.9) and subtract from it the area from the mean to $x = 550$ (Figure 4.10). First the x values must be converted to z numbers so that the normal table may be employed:

$$z_1 = \frac{x_1 - \mu}{\sigma} = \frac{550 - 500}{100} = 0.5$$

$$z_2 = \frac{x_2 - \mu}{\sigma} = \frac{600 - 500}{100} = 1.0$$

$$P(550 < x < 600) = A_{z=1.0} - A_{z=0.5} = 0.3413 - 0.1915 = 0.1498$$

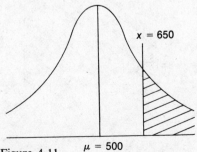

Figure 4.11 $\mu = 500$

(d)   Greater than 650

The table gives areas from the mean to a particular z number. Since the total area under the curve is 1.0 and the curve is symmetrical about the mean, the area on either half is 0.5. (See Figure 4.11.)

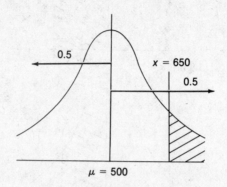

Figure 4.12

Converting $x = 650$ to z, we obtain

$$z = \frac{x - \mu}{\sigma} = \frac{650 - 500}{100} = 1.5$$

(See Figure 4.12.)

$$P(x \geq 650) = \text{area under half the curve} - A_{z=1.5}$$
$$= 0.5 - A_{z=1.5} = 0.5 - 0.4332 = 0.0668$$

(e)   Less than 650 (Figure 4.13)

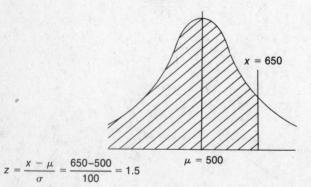

$$z = \frac{x - \mu}{\sigma} = \frac{650 - 500}{100} = 1.5$$

Figure 4.13

There are two generally accepted techniques:

(1) $P(x \leq 650) = 1 - P(x \geq 650) = 1 - 0.0668 = 0.9332$

(2) $P(x \leq 650)$ = area under half the curve + the area from the mean to $x = 0.5 + A_{z=1.5} = 0.5 + 0.4332 = 0.9332$

(f) Less than 375 (Figure 4.14)

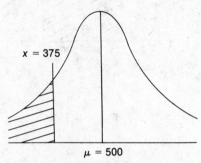

$x = 375$

$\mu = 500$

Figure 4.14

$$z = \frac{x - \mu}{\sigma} = \frac{375-500}{100} = \frac{-125}{100} = -1.25$$

$$P(x < 375) = 0.5 - A_{z=1.25} = 0.5 - 0.3943 = 0.1057$$

Note that the technique used depends on the location of the points with respect to the mean; compare Parts (e) and (f), both "less than" problems. Rather than try and memorize a set of rules, we do well to make a drawing and shade in the portions we are solving for. These drawings plus knowing that the area of the curve is symmetrical and equal to 1, plus Appendix I enables us to solve for any area and, hence, probability, desired.

(g) Between 425 and 640 (Figure 4.15)

[Compare with Parts (a), (b), and (c) to see the effect of the location of the mean.]

$$P(425 < x < 640) = A_{z=0.75} + A_{z=1.4} = 0.2734 + 0.4192 = 0.6926$$

(h) x exactly equals 550

When dealing with the normal distribution, the probability of any exact number is always 0.

If we desire to ascertain the probability that $x \simeq 550$, we build a small increment on either side of 550 and find the area of that increment.

$$z_1 = \frac{549-550}{100} = 0.49 \qquad z_2 = \frac{551-500}{100} = 0.55$$

$$P(549 < x < 551) = A_{z=0.55} - A_{z=0.49} = 0.2088 - 0.1879 = 0.0209$$

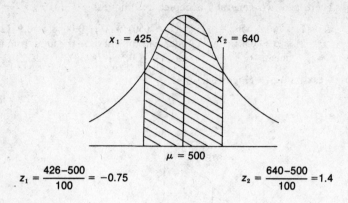

$$z_1 = \frac{426-500}{100} = -0.75 \qquad\qquad z_2 = \frac{640-500}{100} = 1.4$$

Figure 4.15

Example 4.3

Each spring, a fuel tank is emptied and cleaned. In the last 100 gal of the fuel tank, there is always a residue of mercury. The average amount of mercury found over the last 10 years is 20 lb, with a standard deviation of 4 lb. What is the probability of the following, concerning the number of pounds of mercury to be found in the last 100 gal? Let x equal pounds of mercury (Figure 4.16).

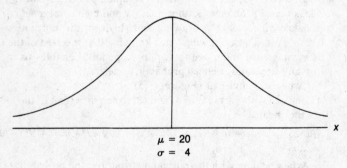

Figure 4.16

(a) Between 16 and 20 lb (Figure 4.17)

The formula $z = (x - \mu)/\sigma$ is used to convert the x value to a z statistic. $z = (16 - 20)/4 = -1$. We look at Appendix I to find the area corresponding to $z = 1$ under the curve from μ to the x value used to generate the z statistic (Figure 4.17). In this part, because

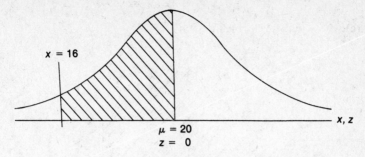

Figure 4.17

the probability sought is between μ and a value, just one step completes the problem. Note that although the negative sign for z indicates that the area is to the left of μ, it has no meaning when we are using the z table.

$$P(16 \leq x < 20) = A_{z=1} = 0.3413$$

(b) Between 15 and 23 lb (Figure 4.18)

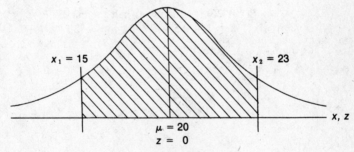

Figure 4.18

The formula $z = (x - \mu)/\sigma$ is used to convert the x values to z statistics:

$$z_1 = \frac{x_1 - \mu}{\sigma} = \frac{15 - 20}{4} = \frac{-5}{4} = -1.25$$

$$z_2 = \frac{x_2 - \mu}{\sigma} = \frac{23 - 20}{4} = \frac{3}{4} = 0.75$$

Using the z statistics in Appendix I, we find the area under the curve from μ to x values. In this part, the areas must be added.

$$P(15 < x < 23) = A_{z=1.25} + A_{z=0.75} = 0.3943 + 0.2734 = 0.6677$$

(c) Less than 26 lb (Figure 4.19)

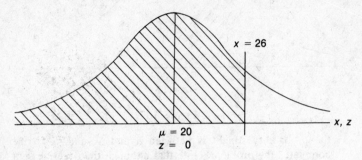

Figure 4.19

$$z = \frac{x - \mu}{\sigma} = \frac{26 - 20}{4} = \frac{6}{4} = 1.5$$

Since the total area under the curve is 1.0, and each half is 0.5, we have

$$P(x < 26) = \text{area under half the curve} + A_{z=1.5}$$
$$= 0.5 + 0.4332 = 0.9332$$

(d) Less than 17 lb (Figure 4.20)

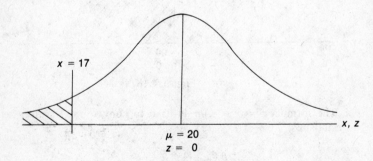

Figure 4.20

$$z = \frac{x - \mu}{\sigma} = \frac{17 - 20}{4} = \frac{-3}{4} = -0.75$$

Using the fact that the area under half the curve is 0.5, we obtain

$$P(x < 17) = \text{area from } x = 0 \text{ to } x = 17$$
$$= 0.5 - \text{area from } x = 17 \text{ to } x = 20$$
$$= 0.5 - A_{z = 0.75} = 0.5 - 0.2734 = 0.2266$$

Part (d) and Part (c) are examples of two different techniques for calculating the probability of x being less than some number, the difference being what side of the mean the point in question is on. Rather than memorizing a set of rules, if one draws the diagram and remembers that the table *always* gives areas from μ to the point in question, then each problem can be reasoned out. The same is true for problems involving values of x greater than some number. The technique for problems involving the probability of x being between two values also depends on whether the points are on the same side of the mean. Again it is preferable to make a drawing rather than to memorize a set of rules. For examples of this, see Parts (a), (b), and the following one, (e).

(e)   Between 22 and 24 lb (Figure 4.21)

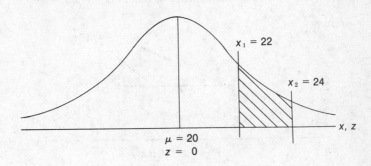

Figure 4.21

$$z_1 = \frac{x_1 - \mu}{\sigma} = \frac{22 - 20}{4} = \frac{2}{4} = 0.5$$

$$z_2 = \frac{x_2 - \mu}{\sigma} = \frac{24 - 20}{4} = 1.0$$

The difference between this part and Parts (a) and (b) is that the points in question are both on the same side of the mean. Appendix I yields the indicated areas for z_1 and z_2 (Figure 4.22). Since the difference between the two areas is desired,

$$P(22 < x < 24) = A_{z = 1.0} - A_{z = 0.5}$$
$$= 0.3413 - 0.1915 = 0.1498$$

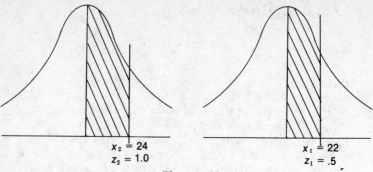

$$x_2 = 24 \qquad\qquad x_1 = 22$$
$$z_2 = 1.0 \qquad\qquad z_1 = .5$$

Figure 4.22

PROBLEMS

Problem 4.1

The average (mean) time spent by commuters on trains from Trenton to New York is 1 h, with a standard deviation of 10 min (Figure 4.23).

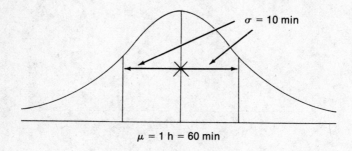

$\sigma = 10$ min

$\mu = 1$ h $= 60$ min

Figure 4.23

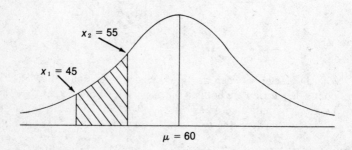

$x_2 = 55$

$x_1 = 45$

$\mu = 60$

Figure 4.24

What is the probability that a commuter will spend the following times on the train.

(a) Between 45 and 55 min
 (See Figure 4.24.)

$$z_1 = \frac{x_1 - \mu}{\sigma} = \frac{45 - 60}{10} = -1.5$$

$$z_2 = \frac{x_2 - \mu}{\sigma} = \frac{55 - 60}{10} = -0.5$$

$$P(45 < x < 55) = A_{z=-15} - A_{z=-0.5}$$

From Appendix I we find

$$A_{z=-1.5} = 0.4332 \qquad A_{z=1.5} = 0.1915$$

$$P(45 < x < 55) = 0.4332 - 0.1915 = 0.2417$$

(b) Between 68 and 70 min
 (See Figure 4.25.)

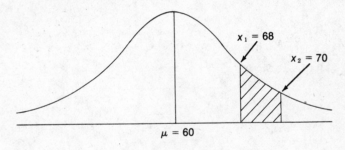

Figure 4.25

$$z_1 = \frac{x_1 - \mu}{\sigma} = \frac{68 - 60}{10} = 0.8$$

$$z_2 = \frac{x_2 - \mu}{\sigma} = \frac{70 - 60}{10} = 1.0$$

$$A_{z=0.8} = 0.2882 \qquad A_{z=1.0} = 0.3413$$

$$P(68 < x < 70) = 0.3413 - 0.2882 = 0.0531$$

(c) Between 56 and 65 min
 (See Figure 4.26.) Note that these two points are on opposite sides of the mean; thus the areas under the curve from the mean to each point must be added instead of subtracted as was done in Parts (a) and (b), where both the points were on the same side of the mean.

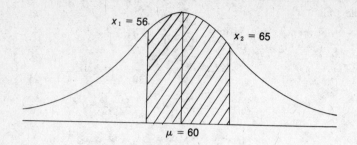

Figure 4.26

$$z_1 = \frac{x_1 - \mu}{\sigma} = \frac{56 - 60}{10} = -0.4$$

$$z_2 = \frac{x_2 - \mu}{\sigma} = \frac{65 - 60}{10} = 0.5$$

$$A_{z=-0.4} = 0.1554 \qquad A_{z=0.5} = 0.1915$$

$$A(56 < x < 65) = 0.1554 + 0.1915 = 0.3469$$

(d) More than 70 min
    (See Figure 4.27.)

$$z = \frac{x - \mu}{\sigma} = \frac{70 - 60}{10} = 1.0$$

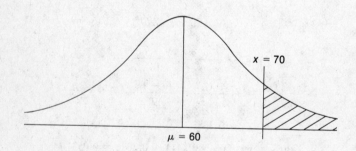

Figure 4.27

$$A_{z=1.0} = 0.3413$$
$$P(x > 70) = 0.5 - 0.3413 = 0.1587$$

(e) Less than 44 min
    (See Figure 4.28.)

$$z = \frac{x - \mu}{\sigma} = \frac{44 - 60}{10}$$

$$z = 1.6$$

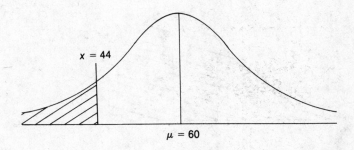

Figure 4.28

$$A_{z=-1.6} = 0.4452$$

$$P(x < 44) = 0.5 - 0.4452 = 0.0548$$

(f) More than 44 min
    (See Figure 4.29.)

$$z = \frac{x - \mu}{\sigma} = \frac{44 - 60}{10}$$

$$z = -1.6$$

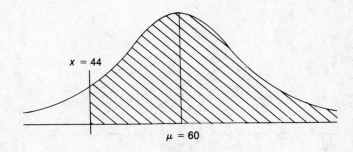

Figure 4.29

$$A_{z=-1.6} = 0.4452$$

$$P(x > 44) = 0.4452 + 0.5 = 0.9452$$

(g) Less than 75 min
(See Figure 4.30.)

$$z = \frac{x - \mu}{\sigma} = \frac{75 - 60}{10} = 1.5$$

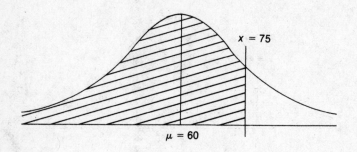

Figure 4.30

$$A_{z=1.5} = 0.4332$$

$$P(x < 75) = 0.4332 + 0.5 = 0.9332$$

(h) Between 55 and 71 min
(See Figure 4.31.)

$$z_1 = \frac{x_1 - \mu}{\sigma} = \frac{55 - 60}{10} = -0.5$$

$$z_2 = \frac{x_2 - \mu}{\sigma} = \frac{71 - 60}{10} = 1.1$$

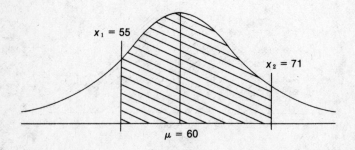

Figure 4.31

$$A_{z_1=0.5} = 0.1915 \qquad A_{z_2=1.1} = 0.3643$$

$$P(55 < x < 71) = 0.1915 + 0.3643 = 0.5558$$

Problem 4.2

Giannini Enterprises buys Indian jewelry that has a mean value per piece of $25, with a standard deviation of $5. Find the probability of the price for a given piece being the following.

(a)   More than $28
    (See Figure 4.32.)

$$z = \frac{x - \mu}{\sigma} = \frac{28 - 25}{5} = 0.6$$

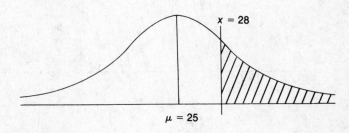

$$\mu = 25$$

Figure 4.32

$$P(x > 28) = 0.5 - A_{z=0.6} = 0.5 - 0.2257 = 0.2743$$

(b)   More than $14
    (See Figure 4.33.)

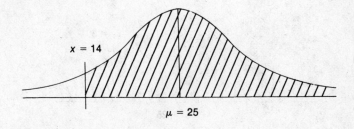

$$\mu = 25$$

Figure 4.33

$$z = \frac{x - \mu}{\sigma} = \frac{14 - 25}{5} = 2.2$$

$$P(x > 14) = 0.5 + A_{z=2.2} = 0.5 + 0.4861 = 0.9861$$

(c) Between \$21 and \$36
     (See Figure 4.34.)

$$z_1 = \frac{x_1 - \mu}{\sigma} = \frac{21 - 25}{5} = -0.8$$

$$z_2 = \frac{x_2 - \mu}{\sigma} = \frac{36 - 25}{5} = 2.2$$

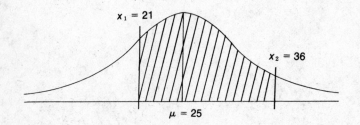

Figure 4.34

$$A_{z_1 = -0.8} = 0.2882 \qquad A_{z_2 = 2.2} = 0.4861$$

$$P(21 < x < 36) = 0.2882 + 0.4861 + 0.7743$$

(d) Between \$30 and \$35
     (See Figure 4.35.)

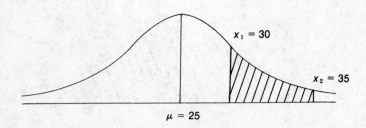

Figure 4.35

$$z_1 = \frac{x_1 - \mu}{\sigma} = \frac{30 - 25}{5} = 1.0$$

$$z_2 = \frac{x_2 - \mu}{\sigma} = \frac{35 - 25}{5} = 2.0$$

$$A_{z_1 = +1.0} = 0.3413 \qquad A_{z=2} = 0.4772$$

$$P(30 < x < 35) = 0.4772 - 0.3413 = 0.1359$$

(e) Between \$4 and \$18
     (See Figure 4.36.)

$$z_1 = \frac{x_1 - \mu}{\sigma} = \frac{4 - 25}{5} = -4.2$$

$$z_2 = \frac{x_2 - \mu}{\sigma} = \frac{18 - 25}{5} = -1.4$$

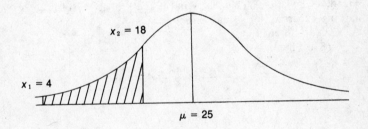

Figure 4.36

Note: When the number is too large for the table, then take the maximum value on the table.

$$A_{z=-4.2} = 0.5 \qquad A_{z=-1.4} = 0.4192$$

$$P(4 < x < 18) = 0.5 - 0.4192 = 0.0808$$

(f) At least \$22
     (See Figure 4.37.)

$$z = \frac{x - \mu}{\sigma} = \frac{22 - 25}{5} = -0.6$$

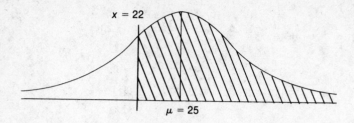

Figure 4.37

$$A_{z=-0.6} = 0.2257$$
$$P(x > 22) = 0.2257 + 0.5 = 0.7257$$

Problem 4.3

Zepke's Plumbing gets an average of 15 calls every 3-day period, with a standard deviation of 2 calls. What is the probability of each of the following occurring in a 3-day period:

(a)   At least 16 calls
       (See Figure 4.38.)

$$z = \frac{x - \mu}{\sigma} = \frac{16 - 15}{2} = 0.5$$

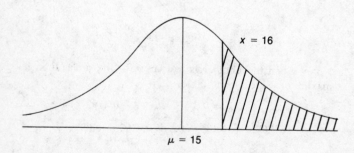

Figure 4.38

$$A_{z=0.5} = 0.1915$$
$$P(x > 16) = 0.5 - 0.1915 = 0.3085$$

(b)   More than 17 calls
      (See Figure 4.39.)

$$z = \frac{x - \mu}{\sigma} = \frac{17 - 15}{2} = 1.0$$

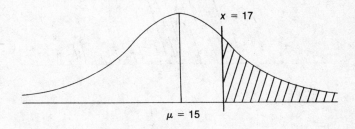

$$x = 17$$

$$\mu = 15$$

Figure 4.39

$$P(x > 17) = 0.5 - A_{z=1.0} = 0.5 - 0.3413 = 0.1587$$

(c)   Less than 18
      (See Figure 4.40.)

$$z = \frac{x - \mu}{\sigma} = \frac{18 - 15}{2} = 1.5$$

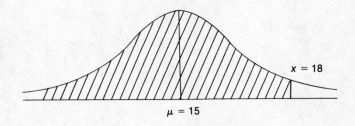

$$x = 18$$

$$\mu = 15$$

Figure 4.40

$$P(x < 18) = 0.5 + A_{z=1.5} = 0.5 + 0.4332 = 0.9332$$

(d)   Less than 25 calls
      (See Figure 4.41.)

$$z = \frac{x - \mu}{\sigma} = \frac{25 - 15}{2} = 5.0$$

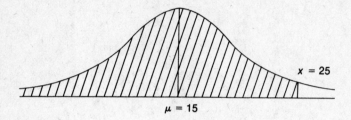

Figure 4.41

$$P(x < 25) = 0.5 + A_{z=5.0} = 0.5 + 0.5 = 1.0$$

Note: This 1.0 is rounded to four decimal places. It really is 0.9999 and something, but to four places it is 1.0.

5

Binomial and Poisson Functions

In building models, the step at which we analyze the data and look for trends or peculiarities of the data is central. Any peculiarities discovered in the data may prove helpful in translating the system to a mathematical one. Researchers in the past have "solved" some systems based on certain specified peculiarities, and if we can legitimately assume that our system has the same peculiarities as one previously solved, then we can apply the results of the previous analysis to our system. The data that are analyzed may be in the form of measurements or previous runs of the same process or previous runs of a similar process or theoretical analysis.

BINOMIAL FUNCTIONS

One specific set of criteria or peculiarities occurs in the binomial model. The characteristics that must be present in the binomial model are the following:

1. Only two possible outcomes—called success and failure—are possible.
2. Probability of success per trial, p, is constant.
3. Probability of failure per trial, $1 - p$, or q, is constant.
4. Trials are independent of each other.

Binomial problems are fairly common, and some of the formulas developed in this chapter may also be used in normal distribution problems that involve proportions, particularly where the issue is whether or not the product being produced is defective or "perfect."

A binomial problem can be concerned with both the success or failure on an individual trial and with the number or probabilities of success or failure in many trials.

Assume that the probability of success per trial is p. If a given occurrence can result in only success or failure, and q is the probability of failure, then $p + q = 1$ and $q = 1 - p$.

Assume that the same trial is to be repeated n times. Then the *mean number of successes in n trials* is equal to the probability of success per trial multiplied by the number of trials, or np.

The *standard deviation* over n trials is $\sqrt{npq}$.

Sometimes the probability of x successes in n trials is desired.

$$\overbrace{\underbrace{p\,p\,p\,p\,p}_{x \text{ successes}} \Big/ \underbrace{q\,q\,q}_{n - x \text{ failures}}}^{n \text{ trials}}$$

One particular way that x successes in n trials could occur would be for the first x trials to be successes and all the remaining trials to be failures, so the number of failures would be the total number of trials minus the number of successes, or $n - x$ failures. If p is the probability of success per independent trial and there are x successes, then the probability of x successes is p multiplied by itself x times, or p^x. The probability of failure is $1 - p$, or q. The probability of $n - x$ failures then is q^{n-x} or $(1 - p)^{n-x}$. Multiplying the probability of x successes by $n - x$ failures gives the probability of one particular arrangement of x successes in n trials as

$$p^x(1 - p)^{n-x}$$

To determine the probability of all ways of getting x successes in n trials, multiply the number of arrangements possible by the probability of one arrangement. The number of arrangements possible can be expressed by the number of combinations possible of n items selected x at a time:

$$\binom{n}{x} = \frac{n!}{(n - x)!\, x!}$$

Recall that the symbol "!" is called factorial and means multiply the number preceding it by all the positive whole numbers between itself and 1.

Combining the two above formulas leads to the following result, which may be applied directly to a problem: the probability of x successes in n trials equals the number of arrangements of n items taken n at a time times the probability of one particular arrangement:

$$\binom{n}{x} p^x(1 - p)^{n-x} = \frac{n!}{(n - x)!\, x!}\, p^x(1 - p)^{n-x}$$

Note that the result is a probability and so it must be between 0 and 1.

Example 5.1

An insurance salesperson sells policies to 5 people, all of identical age and in good health. According to the actuarial tables the probability that a person of this age will be alive in 30 years is ⅔ or 0.67.

(a)   What is the probability that person 1 will be alive in 30 years?

$$P(\text{person } 1) = 0.67$$

(b)   Find the probability that in 30 years exactly 1 person of the 5 will be alive. In this example, $n = 5$, $x = 1$, and $p = 0.67$.

$$P(1) = \binom{5}{1}\left(\frac{2}{3}\right)^1\left(\frac{1}{3}\right)^4 = \frac{5 \cdot 4!}{4! \cdot 1}(0.67)^1(0.33)^4 = 0.04$$

(c)   What is the expected value for the number of years 1 person will live?

$$E(x) = np = 30\left(\frac{2}{3}\right) = 20$$

(d)   What are the variance and standard deviation of the number of years?

$$\text{variance} = npq = (30)(0.67)(0.33) = 6.63$$
$$\text{standard deviation} = \sqrt{\text{variance}} = \sqrt{6.63} = 2.58$$

Example 5.2

30% of the tools produced in a certain manufacturing process turn out to be defective. Find the probability that the following occur in random samples, using the binomial distribution (because a tool is either defective or not).

(a)   Exactly 4 out of 6 are defective. Note that $n = 6$ (6 machines or trials), $x = 4$ (4 are defective), and $p = 0.3$ (30% defective).

$$P(x) = \binom{n}{x}(p)^x(q)^{n-x}$$

$$P(4) = \binom{6}{4}(0.3)^4(0.7)^2$$

$$= \frac{6!}{2! \, 4!}(0.3)^4(0.7)^2$$

$$= \frac{\overset{3}{\cancel{6}} \cdot 5 \cdot \cancel{4!}}{\cancel{2} \cdot \cancel{4!}}(0.0081)(0.49)$$

$$= 0.06$$

(b) There are 4 or 5 defective out of 6 selected. To find the probability of some combination of outcomes the probability of each one must be calculated separately and the results must be added together.

$$P(4 \text{ or } 5) = P(4) + P(5)$$

$$P(4) = 0.06 \text{ [from Part (a)]}$$

$$P(5) = \binom{6}{5}(0.3)^5(0.7)^1 = \frac{6 \cdot 5!}{1 \cdot 5!}(0.0024)(0.7) = 0.01$$

$$P(4 \text{ or } 5) = 0.06 + 0.01 = 0.07$$

Example 5.3

Since a state either passes a law or it does not, the binomial distribution can be applied. The probability that a law on legalized abortion will pass the legislatures of each of 4 southern states is a constant, 0.4. What are the following probabilities?

(a) 3 pass the law.

$$P(3) = \binom{4}{3}(0.4)^3(0.6)^1 = \frac{4 \cdot 3!}{1 \cdot 3!}(0.064)(0.6) = 0.15$$

(b) At least 2 states pass the law.

$$P(\text{at least } 2) = P(2) + P(3) + P(4)$$

$$P(2) = \binom{4}{2}(0.4)^2(0.6)^2 = \frac{\overset{2}{4} \cdot 3 \cdot 2!}{2! \cdot 2!}(0.16)(0.36) = 0.35$$

$$P(3) = 0.15 \text{ [from Part (a)]}$$

$$P(4) = \binom{4}{4}(0.4)^4(0.6)^0 = \frac{4!}{4!}(0.03)(1) = 0.03$$

$$P(\text{at least } 2) = 0.35 + 0.15 + 0.03 = 0.53$$

(c) What is the expected number (value) of states to pass the law?

$$E(x) = np = (0.4) = 1.6$$

This means that if this situation is repeated many times, on the average, 1.6 states would pass the law.

Example 5.4

There are two groups of people who are *independent* of each other. Group A has 3 men and 1 woman while group B has 4 men and 2 women. What is the probability of the following:

(a) Drawing a man from group A on 1 trial?

$$P(\text{man}) = \frac{\text{number of men}}{\text{total number of people}} = \frac{3}{4} = 0.75$$

(b) In 10 trials of drawing someone and then replacing him or her before the next person from group A is drawn, what is the average number of men drawn?

mean number of men = np = (number of trials) (probability per trial)
$$= 10(0.75) = 7.5$$

(c) The standard deviation for drawing 10 times with replacement from group A

$$\text{standard deviation} = \sqrt{npq} = \sqrt{10(0.75)(0.25)} = 1.87$$

Notice that this is the same whether considering men or women, for

$$\sqrt{10(0.75)(0.25)} = \sqrt{10(0.25)(0.75)} = 1.87$$

(d) In 10 trials, the average number of women drawn with replacement from group A

$$\text{mean number of women} = np = 10(0.25) = 2.5$$

(e) A man drawn from group A and a woman from group B? Note the drawings are independent, so

$$P(\text{man A} + \text{woman B}) = P(\text{man A})P(\text{woman B})$$
$$= \frac{3}{4}\left(\frac{2}{6}\right) = 0.25$$

(f) That if 1 person is drawn from each group, the result is a man and a woman? Because it is not specified which group the man or woman must come from, one must consider two possibilities.

$$P(\text{man} + \text{woman}) = P(\text{man A})P(\text{woman B}) + P(\text{woman A})P(\text{man B})$$
$$= \frac{3}{4}\left(\frac{2}{6}\right) + \left(\frac{1}{4}\right)\left(\frac{4}{6}\right) = 0.25 + 0.17 = 0.42$$

(g) That of 2 people drawn from group B, there are no women? We are interested in the number of successes (x) in n trials, where x = number of occurrences of trait concerned with (in this case the number of women) and n is the number of trials (in this case 2) and p is the probability per trial of the occurrence. In this case,

$$\frac{\text{number of women}}{\text{total number of people}} = \frac{2}{6}$$

$$P(x) = \binom{n}{x}(p)^x(1-p)^{n-x} = \frac{n!}{(n-x)!\,x!}(p)^x(1-p)^{n-x}$$

$$P(x=0) = \binom{2}{0}\left(\frac{2}{6}\right)^0\left(1-\frac{2}{6}\right)^{2-0} = \frac{2!}{(2-0)!0!}(1)\left(\frac{4}{6}\right)^2$$

$$= 1(1)\left(\frac{4}{6}\right)^2 = (0.67)^2 = 0.44$$

(Remember: $0! = 1$ and that any number raised to zero power is equal to 1; i.e., $(2/6)^0 = 1$.)

(h) That of 3 people drawn from group B there are at least 2 men

$$P\text{ (at least 2)} = P(2) + P(3)$$
$$n = 3$$
$$p = \frac{4}{6} = 0.67$$
$$x = 2 \text{ and } 3$$
$$P(x) = \binom{n}{x}(p)^x(1-p)^{n-x}$$

for $x = 2$:

$$P(2) = \binom{3}{2}(0.67)^2(1-0.67)^{3-2}$$

$$= \frac{3!}{(3-2)!2!}(0.67)^2(0.33)^1$$

$$= \frac{3\cdot 2\cdot 1}{1\cdot 2\cdot 1}(0.45)(0.33) = 0.45$$

for $x = 3$:

$$P(3) = \binom{3}{3}(0.67)^3(1=0.67)^{3-3}$$

$$= \frac{3!}{3!(3-3)}(0.67)^3(0.33)^0 = 0.30$$

(i) That of 5 people drawn from group B there is at most 1 woman

$$P\text{(at most 1)} = P(0) + P(1)$$
$$n = 5$$

$$p = \frac{2}{6} = 0.33$$

$$x = 0 \text{ and } 1$$

$$P(x) = \binom{n}{x}(p)^x(1 - p)^{n-x}$$

for $x = 0$:

$$P(0) = \binom{5}{0}(0.33)^0(1 - 0.33)^{5-0}$$

$$= \frac{5!}{(5 - 0)!0!}(0.33)^0(0.67)^5 = 1(1)(0.67)^5 = 0.13$$

for $x = 1$:

$$P(1) = \binom{5}{1}(0.33)^1(1 - 0.33)^{5-1}$$

$$= \frac{5!}{(5 - 1)!(1!)}(0.33)(0.67)^4$$

$$= \frac{5 \cdot 4!}{4!}(0.33)(0.67)^4 = 0.33$$

$$P(\text{at most } 1) = P(0) + P(1) = 0.13 + 0.33 = 0.46$$

POISSON DISTRIBUTION

The Poisson distribution is a discrete probability distribution; i.e., it concerns events that occur an exact number of times or not at all. It is generally used to find the probability of some number of successes or failures in a process where the outcomes occur in a random manner so that no pattern is discernible, for example, the number of sneezes in a room in an hour, the number of erasures on a page.

An event can be classified as a "success" or "failure" so long as we are careful to be consistent within the same problem. Since the function is discrete, plugging the appropriate values into the general formula yields the probability of exactly some number of occurrences, so if we wish to ascertain the probability of some combination of events occurring, we must calculate the probability of each event separately and combine the resulting probabilities.

For large samples, in certain circumstances the values obtained from the Poisson distribution approach the values obtained from the normal and binomial distributions. One of the strengths of using the Poisson distribution is that the probability of an event not occurring does not have to be estimated, as it must be for the binomial distribution.

The general formula for the probability of x successes in the Poisson distribution is

$$P(x) = \frac{e^{-\lambda} \lambda^x}{x!}$$

where

e = constant, 2.718
λ = the average number of occurrences in the sample

λ is equal to np, where n = number of trials and p = probability that the event will occur, or the rate of its occurrence per quantity concerned with.

Values for $e^{-\lambda}$

λ	$e^{-\lambda}$	λ	$e^{-\lambda}$	λ	$e^{-\lambda}$
0	1.000	0.7	0.497	5	0.007
0.1	0.905	0.8	0.449	6	0.002
0.2	0.819	0.9	0.407	7	0.0009
0.3	0.741	1	0.368	8	0.0003
0.4	0.670	2	0.135	9	0.0001
0.5	0.607	3	0.050	10	0.00004
0.6	0.549	4	0.018		

Example 5.5

The probability that a certain machine will fail to transcribe a letter correctly is 0.005.

(a) What is the probability that out of 1000 letters transcribed there will be 6 mistakes?

$$\lambda = np$$
$$\lambda = (0.005)(1000) = 5$$

The general form for the Poisson distribution is

$$P(x) = \frac{e^{-\lambda} (\lambda)^x}{x!}$$

Solving the general form for exactly 6 mistakes yields

$$P(6) = \frac{e^{-5}(5)^6}{6!} = \frac{0.007(3125)}{6 \cdot 5 \cdot 4 \cdot 3 \cdot 2 \cdot 1} = 0.03$$

which means that the probability is 0.03, or that 3 times out of 100 trials there will be exactly 6 mistakes in 1000 letters transcribed.

(b) What is the probability that there will be between 6 and 7 mistakes? Because this is a discrete probability distribution, the probability of exactly some number must be calculated, and the results pieced together if necessary.

$$P(\text{between 6 and 7}) = P(6) + P(7)$$
$$P(6) = 0.03 \qquad \text{from Part (a)}$$
$$P(7) = \frac{e^{-5}(5)^7}{7!} = \frac{0.007(15,625)}{7 \cdot 6 \cdot 5 \cdot 4 \cdot 3 \cdot 2 \cdot 1}$$

$$= \frac{109.375}{5040} = 0.02$$

$$P(\text{between 6 and 7}) = 0.03 + 0.02 = 0.05$$

Example 5.6

If there is 1 defect in a piece of cloth 5 ft by 10 ft, what is the probability of there being a defect in a piece of cloth 10 ft by 10 ft? This is an example illustrating that sometimes the given must be converted to a rate per quantity, which is symbolized by λ. The given defective rate is 1 per 50 ft^2; the quantity concerned with is 100 ft.^{2}

$$\lambda = (1/50 \text{ ft}^2)(100 \text{ ft}^2) = 2$$
$$P(x) = \frac{e^{-\lambda}(\lambda)^x}{x!}$$

$$P(1) = \frac{e^{-2}(2)^1}{1!} = \frac{0.135(2)}{1} = 0.27$$

Example 5.7

The number of weekly breakdowns of a computer is a random variable having a Poisson distribution with $\lambda = 0.4$. What is the probability of the following:

(a) There will be 1 breakdown in a week.

$$P(x) = \frac{e^{-\lambda}\lambda^x}{x!}$$

$$P(1) = \frac{e^{-0.4}(0.4)^1}{1!} = \frac{0.670(0.4)}{1} = 0.268$$

(b)   There will be 2 breakdowns in a week.

$$P(x) = \frac{e^{-\lambda} \lambda^x}{x!}$$

$$P(2) = \frac{e^{-0.4}(0.4)^2}{2!} = \frac{0.670(0.4)^2}{2.1} = 0.05$$

(c)   There will not be a breakdown in a week.

$$P(x) = \frac{e^{-\lambda} \lambda^x}{x!}$$

$$P(0) = \frac{e^{-0.4}(0.4)^0}{0!} = \frac{0.670(1)}{1} = 0.67$$

(Remember: Anything, including 0, raised to the zero power is equal to 1.)

(d)   There will be 3 or more breakdowns in a week. There are two approaches to this problem that might be employed.

(1)   Calculate the probabilities of exactly 3, 4, 5, 6, etc., breakdowns and add the results. In using this method, at some point (for example, at $x = 7$) the probabilities will become 0 if rounded to some desired accuracy, and since they will continue to decrease, further calculations will not be necessary.

(2)   Use the complement of the desired result, which is almost always the preferred method when solving for a number greater than some number. The theory is that since the sum of the probabilities of all possible outcomes must equal 1, we have

$$P(0) + P(1) + P(2) + P(3 \text{ or more}) = 1$$
$$P(3 \text{ or more}) = 1 - [P(0) + P(1) + P(2)]$$
$$= 1 - (0.67 + 0.27 + 0.05)$$
$$= 0.01$$

If the probabilities of 0, 1, and 2 outcomes had not previously been generated, they would have had to be in order to solve this question.

(e)   There will be 1 breakdown in 2 wk. Because the base period or quantity involved has changed, the rate for this new quantity must be calculated; hence, λ must be changed. Thus, it is important always to check the quantity involved before starting the calculations.

$$\lambda = (0.4 \text{ breakdowns/wk})(2 \text{ wk}) = 0.8$$

$$P(x) = \frac{e^{-\lambda} \lambda^x}{x!}$$

$$P(1) = \frac{e^{-0.8}(0.8)^1}{1!} = \frac{0.449(0.8)}{1} = 0.36$$

Note, that in comparing this result with Part (a), it becomes obvious that to calculate the probability of a breakdown in 1 wk, we cannot simply divide by 2 the probability of a breakdown in 2 wk.

(f) There will *not* be a breakdown in one-half a week. The base period is not a week, as λ is expressed in the initial data, so λ must be modified to fit a one-half week quantity.

$$\lambda = (0.4 \text{ breakdowns/wk})(0.5 \text{ wk}) = 0.2$$

$$P(x) = \frac{e^{-\lambda} \lambda^x}{x!}$$

$$P(0) = \frac{e^{-0.2}(0.2)^0}{0!} = \frac{0.819(1)}{1} = 0.819$$

(g) There will be *at most* 2 breakdowns in one-half a week.

$$\lambda = (0.4 \text{ breakdowns/wk})(0.5 \text{ wk}) = 0.2$$

$$\lambda = 0.2$$

$$P(\text{at most } 2) = P(0) + P(1) + P(2)$$

$$P(0) = 0.819 \quad \text{from Part (f)}$$

$$P(1) = \frac{e^{-0.2}(0.2)^1}{1!} = \frac{0.819(0.2)}{1} = 0.164$$

$$P(2) = \frac{e^{-2}(0.2)^1}{2!} = \frac{0.819(0.4)}{2 \cdot 1} = 0.016$$

$$P(\text{at most } 2) = 0.819 + 0.164 + 0.016 = 0.999$$

PROBLEMS

Problem 5.1

Two dice are rolled together, with the following outcomes and their probabilities:

Outcome	2	3	4	5	6	7	8	9	10	11	12
Probability	$\frac{1}{36}$	$\frac{2}{36}$	$\frac{3}{36}$	$\frac{4}{36}$	$\frac{5}{36}$	$\frac{6}{36}$	$\frac{5}{36}$	$\frac{4}{36}$	$\frac{3}{36}$	$\frac{2}{36}$	$\frac{1}{36}$

Determine the following:

(a) The probability of a 7 on a roll

$$P(7) = \frac{6}{36} = \frac{1}{6} = 0.167$$

(b) The probability of not rolling a 7

$$P(7^c) = 1 - P(7) = 1 - 0.167 = 0.833$$

(c) The probability of getting a 7 on 1 roll in 3

$$P(x) = \binom{n}{x}(p)^x (1 - p)^{n-x}$$

$$n = 3, \quad x = 1$$

$$P(1) = \binom{3}{1}\left(\frac{1}{6}\right)^1\left(1 - \frac{1}{6}\right)^{3-1} = \frac{3!}{(3 -)!1!} \; \frac{(1)^1}{6} \; \frac{(5)^2}{6}$$

$$= \frac{3 \cdot 2 \cdot 1}{2 \cdot 1 \cdot 1}\left(\frac{1}{6}\right)\left(\frac{25}{36}\right) = 0.348$$

(d) The probability of not getting any 7's in 4 rolls

$$P(0) = \binom{4}{0}\left(\frac{1}{6}\right)^0\left(1 - \frac{1}{6}\right)^{4-0}$$

$$= \frac{4!}{(4 - 0)!0!}\left(\frac{1}{6}\right)^0\left(\frac{5}{6}\right)^4 = \frac{4!}{4!1}\left(\frac{1}{6}\right)^0\left(\frac{5}{6}\right)^4$$

$$= 1(1)\left(\frac{5}{6}\right)^4 = 0.48$$

(e) The probability of getting 5 twice in 3 rolls

$$P(2) = \binom{3}{2}\left(\frac{4}{36}\right)^2\left(1 - \frac{4}{36}\right)^{3-2}$$

$$= \frac{3!}{(3 -)!2!}\left(\frac{4}{36}\right)^2\left(\frac{32}{36}\right)^1$$

$$= \frac{3 \, 2!}{1!2!}\left(\frac{4}{36}\right)^2\left(\frac{32}{36}\right) = 3\left(\frac{4}{36}\right)^2 \frac{32}{36}$$

$$= 0.033$$

(f) The probability of getting 5 at least twice in 3 rolls

$$P(\text{at least } 2) = P(2) + P(3)$$

$$P(2) = 0.033 \qquad \text{from Part (e)}$$

$$P(3) = \binom{3}{3}\left(\frac{4}{36}\right)^3\left(1 - \frac{4}{36}\right)^{3-3}$$

$$= \frac{3!}{(3-3)!0!}\left(\frac{4}{36}\right)^3(1) = \left(\frac{4}{36}\right)^3 = 0.001$$

$$P(\text{at least } 2) = P(2) + P(3) = 0.033 + 0.001 = 0.034$$

(g) The probability of getting a 7 or an 11 on 2 rolls out of 3

$$n = 3, \qquad x = 2$$

$$P = P(7) + P(11) = \frac{6}{36} + \frac{2}{36} = \frac{8}{36}$$

$$P(2) = \binom{3}{2}\left(\frac{8}{36}\right)^2\left(1 - \frac{8}{36}\right)^{3-2}$$

$$= \frac{3!}{(3-2)!2!}\left(\frac{8}{36}\right)^2\left(\frac{28}{36}\right)^1$$

$$= \frac{3 \cdot 2!}{1 \cdot 2!}\left(\frac{8}{36}\right)\frac{28}{36} = 3(0.05)(0.77)$$

$$= 0.116$$

(h) The probability of obtaining at least a 10 on 3 out of 5 rolls

For a given roll

$$P(10 \text{ or more on a given roll}) = P(10) + P(11) + P(12)$$

$$= \frac{3}{36} + \frac{2}{36} + \frac{1}{36} = \frac{1}{6}$$

For 3 out of 5 rolls

$$n = 5, \qquad x = 3, \qquad \text{and} \qquad p = \frac{1}{6}$$

$$P(3) = \binom{5}{3}\left(\frac{1}{6}\right)^3\left(1 - \frac{1}{6}\right)^{5-3}$$

$$= \frac{5!}{(5-3)!3!}\left(\frac{1}{6}\right)^3\left(\frac{5}{6}\right)^2 = 10(0.005)(0.69)$$

$$P(\text{at least } 10 \text{ on 3 out of 5 rolls}) = 0.035$$

(i) The probability of a 12 on 1 out of 2 rolls

$$P(1) = \binom{2}{1}\left(\frac{1}{36}\right)^1\left(1 - \frac{1}{36}\right)^{2-1}$$

$$= \frac{2!}{(2 - 1)!1!} \left(\frac{1}{36}\right) \frac{35}{36}$$

$$= \frac{2 \cdot 1!}{1!1!} \left(\frac{1}{36}\right) \frac{35}{36}$$

$$P(1) = 0.054$$

Problem 5.2

A manufacturer of batteries claims that the probability that one of her batteries operates for 6 mo is 90%. Mr. Davis buys 50 of these batteries.

(a) What is the expected value for the number of batteries that will last Mr. Davis 6 months?

$E(x) = np = (50 \text{ batteries})(0.90) = 45 \text{ batteries}$

(b) What is the probability that in 6 mo all 5 of the batteries sold yesterday will be operating?

$$n = 5, \qquad x = 5, \qquad p = 0.90$$

$$P(x) = \binom{n}{x}(p)^x(1 - p)^{n-x}$$

$$P(5) = \binom{5}{5}(0.90)^5(1 - 0.90)^{5-5}$$

$$= \frac{5!}{(5 - 5)!5!}(0.90)^5(0.10)^0 = 1(0.90)^5 \, 1$$

$$= 0.59$$

(c) What is the probability that at least 3 of the 5 batteries sold will be operating in 6 mo (do two different ways)?

(1)
$$P(\text{at least } 3) = P(3) + P(4) + P(5)$$

$$P(x) = \binom{n}{x}(p)^x \, (1 - p)^{n-x}$$

$$n = 5, \qquad p = 0.9, \qquad x = 3, 4, 5$$

$$P(3) = \binom{5}{3}(0.9)^3 \, (1 - 0.9)^{5-3} = \frac{5!}{(5 - 3)!3!}(0.9)^3(0.1)^2$$

$$= \frac{5 \cdot 4 \cdot 3!}{2 \cdot 1 \cdot 3!}(0.9)^3(0.1)^2 = 0.073$$

$$P(4) = \binom{5}{4}(0.9)^4(1 - 0.9)^{5-4} = \frac{5!}{(5 - 4)!4!} (0.9)^4(0.1)^1$$

$$= \frac{5 \cdot 4!}{1 \cdot 4!} (0.9)^4(0.1) = 0.328$$

$$P(5) = \binom{5}{5}(0.9)^5(1 - 0.9)^{5-5} = \frac{5!}{(5 - 5)!5!} (0.9)^5(0.1)^0$$

$$= \frac{5!}{0!5!} (0.9)^5 1 = 1(0.9)^5 1 = 0.590$$

$P(\text{at least } 3) = P(3) + P(4) + P(5) = 0.073 + 0.328 + 0.590$
$$= 0.991$$

(2)

$P(\text{at least } 3) = 1 - P(\text{less than } 3)$

$$= 1 - [P(0) + P(1) + P(2)]$$

$$P(x) = \binom{n}{x}(p)^x (1 - p)^{n-x}$$

$$n = 5, \quad p = 0.9, \quad x = 0, 1, 2$$

$$P(0) = \binom{5}{0}(0.9)^0 (1 - 0.9)^{5-0}$$

$$= \frac{5!}{(5 - 0)!0!} (0.9)^0 (0.1)^5 = \frac{5!}{5!0!} (1)(0.1)^5$$

$$= 1(1)(0.1)^5 = 0.00001$$

$$P(1) = \binom{5}{1}(0.9)^1 (1 - 0.9)^{5-1} = \frac{(0.9)^1(0.1)^4 \, 5!}{(5 - 1)!1!}$$

$$= \frac{5 \cdot 4!}{4!1} (0.9)(0.1)^4 = 0.00045$$

$$P(2) = \binom{5}{2}(0.9)^2 (1 - 0.9)^{5-2} = \frac{5!}{(5 - 2)!2!} (0.9)^2(0.1)^3$$

$$= \frac{5 \cdot 4^2 \cdot 3!}{3! \cdot 2 \cdot 1} (0.9)^2 (0.1)^3 = 10(0.81) (0.001)$$

$$= 0.0081$$

$P(\text{at least } 3) = 1 - [P(0) + P(1) + P(2)]$
$$= 1 - (0.00001 + 0.00045 + 0.0081)$$

$$= 1 - (0.00856) \text{ round off to } 0.009$$
$$= 1 - (0.009) = 0.991$$

Note that the answer is the same by either method. In the second method, the numbers were carried to more decimal places than usually is done in order to illustrate this point.

(d) What is the variance of the life of Mr. Davis' batteries?

$$\text{variance} = npq = np(1 - p) = 50 \,(0.9)(0.1) = 4.5$$

(e) What is the standard deviation of the life of Mr. Davis' batteries?

$$\text{standard deviation} = \sqrt{\text{variance}} = \sqrt{4.5} = 2.12 \text{ mo}$$

Problem 5.3

The probability of a pregnant woman giving birth to a boy is 0.5 and to a girl is 0.5, with multiple births not considered.

(a) What is the probability of one's first child being a girl?

$$P(g) = 0.5$$

(b) What is the probability of having 2 boys in a row?

$$P(x) = \binom{n}{x}(p)^x \,(1 - p)^{n-x}$$

$$n = 2, \qquad x = 2, \qquad p = 0.5$$

$$P(2) = \binom{2}{2}(0.5)^2(1 - 0.5)^{2-2} = \frac{2!}{(2 - 2)!2!} \,(0.5)^2(0.5)^0$$

$$= 1(0.5)^2 \, 1 = 0.25$$

(c) What is the expected number of boys in 30 births?

$$E(x) = np = 30(0.5) = 15$$

(d) What is the variance for the number of boys in 30 births?

$$\text{variance} = npq = np(1 - p) = 30(0.5)(1 - 0.5) = 30(0.5)(0.5)$$
$$= 7.5$$

(e) What is the standard deviation for the birth of 30 boys?

$$\text{standard deviation} = \sqrt{\text{variance}} = \sqrt{7.5} = 2.74$$

(f) What is the probability of having 4 girls in a row?

$$P(4) = \binom{4}{4}(0.5)^4 \,(1 - 0.5)^{4-4}$$

$$= \frac{4!}{(4-4)!4!} \, (0.5)^4(0.5)^0 = 1(0.5)^4 \, 1$$

$$= 0.0625$$

Problem 5.4

Given a Poisson function with $\lambda = 0.6$, determine the probability of the following:

(a) $x = 2$

for Poisson

$$P(x) = \frac{e^{-\lambda} \lambda^x}{x!}$$

$$P(2) = \frac{e^{-0.6}(0.6)^2}{2 \cdot 1} = \frac{0.549(0.6)^2}{2 \cdot 1} = 0.099$$

(b) $x = 4$

$$P(4) = \frac{e^{-0.6}(0.6)^4}{4!} = \frac{0.549(0.6)^4}{4 \cdot 3 \cdot 2 \cdot 1} = 0.003$$

(c) $x = 0$

$$P(0) = \frac{e^{-0.6}(0.6)^0}{0!} = \frac{0.549(0.6)^0}{1} = .549$$

(d) x is greater than or equal to 1 and less than or equal to 3, or $1 \le x \le 3$

The Poisson is a discrete function, so

$$P(1 \le x \le 3) = P(1) + P(2) + P(3)$$

$$P(1) = \frac{e^{-0.6}(0.6)^1}{1!} = \frac{0.549(0.6)}{1} = 0.329$$

$$P(2) = \frac{e^{-0.6}(0.6)^2}{2!} = \frac{0.549(0.6)^2}{2 \cdot 1} = 0.099$$

$$P(3) = \frac{e^{-0.6}(0.6)^3}{3!} = \frac{0.549(0.6)^2}{3 \cdot 2 \cdot 1} = 0.020$$

$$P(1 \le x \le 3) = 0.329 + 0.099 + 0.020$$

$$= 0.448$$

(e) x is greater than 1 and less than 3, or $1 < x < 3$

Here the endpoints (1 and 3) are *not* included, and since the

Poisson can only be solved for integer values of x (that is, whole numbers), we have

$$P(1 < x < 3) = P(2) = 0.099 \text{ from Part (d)}$$

(f) $x \geq 2$

Since solving this directly would require the summation of $P(2) + P(3) + P(4) + \cdots + P(n)$, it is simpler to use the complement relationship present in all probability functions, or $P(x \geq 2) = 1 - P(x < 2)$:

$$P(x \geq 2) = 1 - P(x < 2) = 1 = [P(0) + P(1)]$$

$$P(0) = 0.549 \text{ from Part (c)}$$

$$P(1) = 0.329 \text{ from Part (d)}$$

$$P(x \geq 2) = 1 - (0.549 + 0.329) = 0.122$$

Problem 5.5

Given a Poisson function with $\lambda = 0.8$, determine the probability of the following:

(a) $x = 0$

$$P(0) = \frac{e^{-0.8}(0.8)^0}{0!} = \frac{0.449(0.8)^0}{0!} = \frac{0.449(1)}{1} = 0.449$$

(b) $x = 6$

$$P(6) = \frac{e^{-0.8}(0.8)^6}{6!} = \frac{0.449(0.8)^6}{6 \cdot 5 \cdot 4 \cdot 3 \cdot 2 \cdot 1} = \frac{0.449(0.262)}{720} = 0.0002$$

(c) $x = 3$

$$P(3) = \frac{e^{-0.8}(0.8)^3}{3!} = \frac{0.449(0.8)^3}{3 \cdot 2 \cdot 1} = 0.038$$

(d) $x \leq 2$

$$P(x \leq 2) = P(0) + P(1) + P(2)$$

$$P(0) = 0.449 \text{ from Part (a)}$$

$$P(1) = \frac{e^{-0.8}(0.8)^1}{1!} = \frac{0.449(0.8)}{1} = 0.359$$

$$P(2) = \frac{e^{-0.8}(0.8)^2}{2!} = \frac{0.449(0.8)^2}{2 \cdot 1} = 0.144$$

$$P(x \leq 2) = 0.449 + 0.359 + 0.144 = 0.952$$

(e) $x > 2$

$P(x > 2) = 1 - P(x \leq 2) = 1 - 0.952 = 0.048$ from Part (d)

Problem 5.6

George discovers that there is a probability of 0.25 that a customer will leave his restaurant in any given hour without paying the bill. He is concerned with 2-h increments. Determine the probability of the following:

(a) That no one will leave without paying the bill $(x = 0)$

$\lambda = 0.25(2) = 0.5$ for a 2-h period

$$P(0) = \frac{e^{-0.5}(0.5)^0}{0!} = \frac{0.607(0.5)^0}{0!} = \frac{0.607(1)}{1} = 0.607$$

(b) That 1 person will leave without paying the bill

$$P(1) = \frac{e^{-0.5}(0.5)^1}{1!} = \frac{0.607(0.5)}{1} = 3.04$$

(c) That more than 1 but less than 3 will leave without paying the bill, or $1 < x < 3$

The endpoints of 1 and 3 are not included, so

$$P(1 < x < 3) = P(2) = \frac{e^{-0.5}(0.5)^2}{2!} = \frac{0.607(0.5)^2}{2 \cdot 1} = 0.076$$

(d) That between 1 and 3 will leave without paying the bill, or $1 \leq x \leq 3$

The endpoints are included, so

$P(1 \leq x \leq 3) = P(1) + P(2) + P(3)$

$P(1) = 0.304$ from Part (b)

$P(2) = 0.076$ from Part (c)

$$P(3) = \frac{e^{-0.5}(0.5)^3}{3!} = \frac{0.607(0.5)^3}{3 \cdot 2 \cdot 1} = 0.013$$

$P(1 \leq x \leq 3) = P(1) + P(2) + P(3) = 0.304 + 0.076 + 0.013$
$= 0.393$

(e) That at least 3 will leave without paying their bills

$P(x \geq 3) = 1 - [P(0) + P(1) + P(2)]$

$P(0) = 0.607$ from Part (a)

$P(1) = 0.304$ from Part (b)

$P(2) = 0.076$ from Part (c)

$$P(x \geq 3) = 1 - (0.607 + 0.304 + 0.076) = 1 - 0.987$$
$$= 0.01$$

Problem 5.7

George, from the preceding example, discovers that the true value of λ is 5, instead of 0.5, as he previously thought. Based on $\lambda = 5$, determine the following probabilities:

(a) $x = 0$

$$P(0) = \frac{e^{-\lambda} \lambda^x}{x!} = \frac{e^{-5}(5)}{0!} = 0.007$$

(b) $x = 1$

$$P(1) = \frac{e^{-5}(5)^1}{1!} = \frac{0.007(5)}{1} = 0.035$$

(c) $1 < x < 3$

$$P(1 < x < 3) = P(2) = \frac{e^{-5}(5)^2}{2!} = \frac{0.007(25)}{2(1)} = 0.088$$

(d) $1 \leq x \leq 3$

$$P(1 \leq x \leq 3) = P(1) + P(2) + P(3)$$
$$P(1) = 0.035 \text{ from Part (b)}$$
$$P(2) = 0.088 \text{ from Part (c)}$$
$$P(3) = \frac{e^{-5}(5)^3}{3!} = \frac{0.007(125)}{3 \cdot 2 \cdot 1} = 0.146$$
$$P(1 \leq x \leq 3) = 0.035 + 0.088 + 0.146 = 0.269$$

(e) $x \geq 3$

$$P(x \geq 3) = 1 - [P(0) + P(1) + P(2)]$$
$$P(0) = 0.007 \text{ from Part (a)}$$
$$P(1) = 0.035 \text{ from Part (b)}$$
$$P(2) = 0.088 \text{ from Part (c)}$$
$$P(x \geq 3) = 1 - (0.007 + 0.035 + 0.088) = 1 - 0.13 = 0.87$$

Note: Compare the effect of the size of λ—the preceding two examples are identical except that the value of λ in Problem 5.7 was 10 times greater than the λ used in Problem 5.6. Note that the corresponding probabilities are *not* exactly 10 times larger than in the previous problem.

6

Statistical Inferences

Inferences are statements that are made about a large entity based on information about a smaller entity. If the inferences are based on statistical principles, they are referred to as *statistical inferences*.

Statistical inferences about characteristics of an entire class of objects (the *population*) may be made on the basis of data collected about a portion of the population (the *sample*). In the model building process outlined in Chapter 1, the data are of characteristics of the sample. If it is determined by the analysis that the distribution of the data is approximately normal, then the model is based on the normal distribution. The equations for this model have previously been generated based on the assumption that the data conform to a normal distribution. The sample data are plugged into the equations, and the calculations lead to numerical predictions concerning the population.

The analysis presented in this chapter is applicable for samples of 30 or more. Chapter 8 contains the analysis that should be used for problems involving samples of less than 30.

ACTUAL DATA

The formulae utilized within this chapter are generated by algebraic manipulations of the general form of a normal equation, which is

$$z = \frac{\bar{x} - \mu}{\sigma/\sqrt{N}} \tag{1}$$

Once sample data are known, a projection about the population can be made by the normal distribution, which we transform into

$$\mu = \bar{x} \pm z \frac{\sigma}{\sqrt{N}} \tag{2}$$

Formula (2) shows that μ (the population mean) can exactly be estimated by $\bar{x}$ (the sample mean), with an error term of $z\, \sigma/\sqrt{N}$, where N is the sample size, σ is the population standard deviation, and z is derived from the desired confidence level in the result. The total of the confidence level plus the level of significance equals 1.

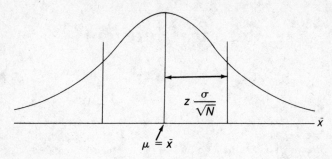

$$z \frac{\sigma}{\sqrt{N}}$$

$$\mu = \bar{x}$$

Figure 6.1

A problem arises because σ is needed for a determination of μ, so a substitution of $\bar{\sigma}$ (sample standard deviation) for σ (population standard deviation) is made. Hence an estimate for μ can be made based strictly on sample data: $\bar{x}, \bar{\sigma}, N$, and a confidence level z that is specified with the problem or that can be extracted from a normal table (see Appendix IV) if the percentage of confidence in the result is specified.

$$\mu = \bar{x} \pm z \frac{\bar{\sigma}}{\sqrt{N}} \tag{3}$$

PROPORTIONS

If the data are given in proportions, then by substituting p for $\bar{x}$ and $\sqrt{p(1-p)}$ for σ in Formula (3), we obtain

$$p = \bar{p} \pm z \sqrt{\frac{\bar{p}(1-\bar{p})}{N}} \tag{4}$$

where $\bar{p}$ is sample proportion possessing the desired trait and p is population proportion possessing the desired trait.

The desired error term can be specified and calculations made to determine the minimum sample size that will keep the error term at less than the specified quantity. Note that the error term decreases as the sample size increases.

SAMPLE SIZE

The above formulas, listed with the z statistic, are used if the sample size is greater than 30. If the sample size is less than 30, then see Chapter

9, "Difference Between Means," section on "t Statistic" for a technique that is very similar.

Example 6.1

The average weight for 36 items sampled was 22 oz, with a standard deviation of 5 oz.

(a) Estimate the average weight for all the items. The best single estimate of the population mean is the sample mean, so

$$\text{population mean } \mu = \bar{x} = 22 \text{ oz}$$

(b) Estimate with 95% confidence the population mean.

$$\mu = \bar{x} \pm z \, \frac{\bar{\sigma}}{\sqrt{n}} = 22 \pm 1.96 \, \frac{5}{\sqrt{36}}$$

$$= 22 \pm 1.63 \text{ oz}$$

The population mean, therefore, should lie between $22 - 1.63$ and $22 + 1.63$, or $20.37 \leq \mu \leq 23.63$ oz, and you are 95% confident of this.

Example 6.2

20% of a sample of 100 ft^3 of Knoxville water is polluted. Find the 90% confidence limits for the proportion of all the polluted water disposed.

$$p = \bar{p} \pm z \sqrt{\frac{\bar{p}(1 - \bar{p})}{N}}$$

$$= 0.20 \pm 1.645 \sqrt{\frac{0.20(0.80)}{100}} = 0.20 \pm 0.07$$

or $0.13 \leq p \leq 0.27$, so between 13% and 27% of the water was polluted.

Example 6.3

A candidate for local office wants to determine how popular she is, so she takes a sample and finds 52 people favoring her out of 100 asked, with a variance of 81. How large a sample should she select in order to be 95% confident that the error in the estimated population mean will not exceed 8 votes?

$$\text{error term} \leq 8 \text{ votes}$$

$$\text{so } z\,\frac{\bar{\sigma}}{\sqrt{N}} \leq 8$$

Make an equality and solve, which will yield a minimum value:

$$z\,\frac{\bar{\sigma}}{\sqrt{N}} = 8$$

$$\sqrt{N} = \frac{z\bar{\sigma}}{8}$$

$$N = \left(\frac{z\bar{\sigma}}{8}\right)^2$$

$$N = \left(\frac{1.96(9)}{8}\right)^2 \cong 5$$

She must sample at least 5 people.

Example 6.4

To demonstrate the effect of the magnitude of the error term, hold the other terms constant and reduce the desired error term to 2 votes.

$$N = \left(\frac{z\bar{\sigma}}{2}\right)^2 = \left(\frac{1.96(9)}{2}\right)^2 \simeq 78$$

The minimum sample size to hold the error term to 2 votes is 78.

Comparing this example with the previous one demonstrates that one way to reduce the error term is to take larger samples. As one would logically expect, the result becomes more precise as the size of the sample is increased.

PROBLEMS

Problem 6.1

A sample of paint cans is taken by inspector Todd, who finds 12 defective cans out of the 64 cans he inspected. What should Todd's estimates for the following be?

(a) What is the proportion of defective cans in the population from which the cans were selected?

$$p = \bar{p} = \frac{\text{number defective}}{\text{number in sample}} = \frac{12}{64} = 0.19$$

(b) With 95% confidence, what would his estimate be for the proportion of defective ones in the population?

$$p = \bar{p} \pm z \sqrt{\frac{\bar{p}(1 - \bar{p})}{N}} = \frac{12}{64} \pm 1.96 \sqrt{\frac{\frac{12}{64}\left(1 - \frac{12}{64}\right)}{64}}$$

$$= 0.19 \pm 1.96 \sqrt{\frac{0.19(1 - 0.19)}{64}} = 0.19 \pm 1.96 \sqrt{0.0024}$$

$$= 0.19 \pm 1.96(0.049) = 0.19 \pm 0.096$$

or $0.094 < p < 0.286$

(c) With 95% confidence, what would his estimate be for the proportion of *non*defective ones?

$$p = \bar{p} \pm z \sqrt{\frac{\bar{p}(1 - \bar{p})}{N}} = \frac{52}{64} \pm 1.96 \sqrt{\frac{\frac{52}{64}\left(1 - \frac{52}{64}\right)}{64}}$$

$$= 0.81 \pm 1.96 \sqrt{\frac{0.81(0.19)}{64}}$$

$$= 0.81 \pm 0.096$$

or $0.714 < p < 0.906$

Problem 6.2

Sonia, a statistician, takes measurements for the number of ounces that are actually in each bottle of a soft drink and obtains the following sample:

Number of ounces	Numbers of bottles containing these ounces
13	5
12	20
11	16
10	14
9	8
8	2
Total	65

(a) What is the sample mean?

$$\bar{x} = \sum_{i=1}^{n} \frac{x_i f(x_i)}{x_i \, Nf(x_i)} = \frac{1}{N} \sum x_i f(x_i)$$

$$= \frac{1}{65} (709) = 10.9, \text{ or approximately 11 oz/bottle}$$

(b)   What is the sample variance?

$$\bar{\sigma}^2 = \sum_{i=1}^{n} (x_i - \bar{x})^2 \, \frac{f(x_i)}{N} = \frac{1}{N} \Sigma (x - \bar{x})^2 f(x_i)$$

$$= \frac{1}{65}[(13 - 11)^2 \, 5 + (12 - 11)^2 \, 20 + (11 - 11)^2 \, 16$$

$$+ (10 - 11)^2 \, 14 + (9 - 11)^2 \, 8 + (8 - 11)^2 \, 2]$$

$$= \frac{1}{65} (20 + 20 + 0 + 14 + 32 + 18)$$

$$= \frac{104}{65}$$

$$= 1.6$$

(c)   What is the sample standard deviation?

$$\bar{\sigma} = \sqrt{\bar{\sigma}^2} = \sqrt{1.6} = 1.26 \text{ oz/bottle}$$

(d)   What is the best point estimate for the population mean?

$$\mu = \bar{x} = 10.9 \text{ oz/bottle}$$

(e)   What should Sonia's estimate be for the 95% interval for the number of ounces per bottle for all bottles?

$$\mu = \bar{x} \pm z \, \frac{\bar{\sigma}}{\sqrt{N}} = 10.9 \pm 1.96 \, \frac{1.26}{\sqrt{65}}$$

$$= 10.9 \pm 0.31 \text{ oz/bottle}$$

or $10.59 < \mu < 11.21$ oz/bottle

(f)   What should be Sonia's estimate for the 99% interval for the number of ounces per bottle for all bottles?

$$\mu = \bar{x} \pm z \, \frac{\bar{\sigma}}{\sqrt{N}} = 10.9 \pm 2.57 \, \frac{1.26}{\sqrt{65}}$$

$$= 10.9 \pm 0.40 \text{ oz/bottle}$$

or $10.5 < \mu < 11.3$ oz/bottle

(g) What should Sonia's estimate be for the 70% interval for the number of ounces per bottle for all bottles?

$$\mu = \bar{x} \pm z \, \frac{\bar{\sigma}}{\sqrt{N}} = 10.9 \pm 1.04 \, \frac{1.26}{\sqrt{65}}$$

$$= 10.9 \pm 0.16 \text{ oz/bottle}$$

or $\qquad 10.74 < \mu < 11.06$ oz/bottle

(h) What is the effect of increasing the confidence level? Compare the above examples to see that in order to be more confident of μ lying within a specified area, we should increase the area.

Problem 6.3

Michele, a state official, wants to ascertain how many items have been raised in price at the grocery store, so she obtains a sample that shows that prices of 65 items out of 90 have been raised, with a variance of 25. How large a sample should she select in order to be 95% confident that the error in the estimated population mean will not exceed 4 items?

To estimate the population mean $\mu = \bar{x} \pm z \, \dfrac{\bar{\sigma}}{\sqrt{N}}$

$\bar{x}$ is the point estimate for μ and $z \, \dfrac{\bar{\sigma}}{\sqrt{N}}$ is the "error" term (which in this problem must be less than or equal to 4), so we make an equality and solve for N:

$$z \, \frac{\bar{\sigma}}{\sqrt{N}} \leq 4$$

We make the inequality into an equality so that the answer is the minimum number of items that Michele should sample to obtain the desired results:

$$\sqrt{N} = \frac{z\sigma}{4}$$

$$N = \left(\frac{z\bar{\sigma}}{4} \right)^2$$

for 95% ($z = 1.96$)

$$N = \left(\frac{1.96 \sqrt{25}}{4} \right)^2 = \left(\frac{1.96(5)}{4} \right)^2 = 6$$

Problem 6.4

Jill, a student, misses 8 out of 40 questions on one part of an exam. Concerning the entire exam, what are the best estimates for the following:

(a) The proportion of wrong answers for the entire exam

$$p = \bar{p} = \frac{\text{number of wrong answers}}{N} = \frac{8}{40} = 0.20$$

(b) The best point estimate for the mean number of wrong answers on an exam of 200 questions

$$\mu = N\bar{p} = 200 \left(\frac{8}{40} \right) = 40$$

(c) The proportion of correct answers

$$p = \bar{p} = \frac{\text{number of correct answers}}{N} = \frac{32}{40} = 0.80$$

(d) The best estimate, with 95% confidence, for the percentage that would be missed on the entire exam

$$p = \bar{p} \pm z \sqrt{\frac{\bar{p}(1 - \bar{p})}{N}} = \frac{8}{40} \pm 1.96 \sqrt{\frac{\frac{8}{40}\left(1 - \frac{8}{40}\right)}{40}}$$

$$= \frac{8}{40} \pm 1.96 \sqrt{\frac{0.2(0.8)}{40}} = 0.2 \pm 0.124 \text{ of the questions that}$$

would be missed

or   7.6% < p < 32.4%

(e) The 90% confidence estimates for the number *not* missed on the entire exam

$$p = \bar{p} \pm z \sqrt{\frac{\bar{p}(1 - \bar{p})}{N}} = \frac{32}{40} \pm 1.645 \sqrt{\frac{\frac{32}{40}\left(1 - \frac{32}{40}\right)}{40}}$$

$$= 0.8 \pm 1.645 \sqrt{\frac{0.8(0.2)}{40}} = 0.8 \pm 0.104$$

or   69.6% < p < 90.4%

Problem 6.5

Chris, a cost accountant, samples the expense accounts of the sales-persons of his company, who all eat at the same restaurant, and arrives at the following data:

Dollars spent	Number of persons
2	8
3	15
4	20
5	10
6	6
7	5
Total	64

(a) What is the sample mean?

$$\bar{x} = \frac{1}{N} \sum_{i=1}^{n} x_i f(x)_i$$

$$= \frac{1}{64} [2(\$8) + 3(\$15) + 4(\$20) + 5(\$10) + 6(\$6) + 7(\$5)]$$

$$= \frac{\$262}{64} = \$4.09/\text{lunch}$$

(b) What is the sample variance? (Use $\bar{x} = \$4/\text{lunch}$ to simplify the calculations.)

$$\bar{\sigma}^2 = \frac{1}{N} \sum_{i=1}^{n} (x_i - \bar{x})^2 f(x_i)$$

$$= \frac{1}{64} [(2 - 4)^2\, 8 + (3 - 4)^2\, 15 + (4 - 4)^2\, 20 + (5 - 4)^2\, 10$$

$$+ (6 - 4)^2\, 6 + (7 - 4)^2\, 5]$$

$$= \frac{1}{64} (32 + 15 + 0 + 10 + 24 + 45)$$

$$= \frac{126}{64} = 1.97$$

(c) What is the sample standard deviation?

$$\bar{\sigma} = \sqrt{\text{variance}} = \sqrt{\bar{\sigma}^2} = \sqrt{1.97} = \$1.4/\text{lunch}$$

(d) What is the 90% interval for the number of dollars spent per lunch for the population of all Chris's salespersons?

$$\mu = \bar{x} \pm z \, \frac{\bar{\sigma}}{\sqrt{N}} = 4 \pm 1.645 \, \frac{1.4}{\sqrt{64}} = \$4 \pm \$0.29/\text{lunch}$$

or the average cost for lunch is between $3.71 and $4.29.

(e) What is the 60% interval for the number of dollars spent per lunch for the population of all Chris's salespersons?

$$\mu = \bar{x} \pm x \, \frac{\bar{\sigma}}{\sqrt{N}} = 4 \pm 0.84 \, \frac{1.4}{\sqrt{64}} = \$4 \pm \$0.15/\text{lunch}$$

or the average lunch costs between $3.85 and $4.15.

Problem 6.6

To illustrate the importance and effect of selecting a representative sample, take the data for the preceding example and add 16 salespersons who eat heavily and average $20/lunch. In the parts below, compare the answers from this example with those calculated in the previous example.

(a) What is the sample mean?

$$\bar{x} = \frac{1}{80} \, [2(\$8) + 3(\$15) + 4(\$20) + 5(\$10) + 6(\$6) + 7(\$5) + 20(\$16)]$$

$$= \frac{1}{80} \, \$582 = \$7.3/\text{lunch}$$

(b) What is the sample variance?

$$\bar{\sigma}^2 = \frac{1}{N} \sum_{i=1}^{n} (x_i - \bar{x}) f(x_i) = \frac{1}{80} \Sigma (x - 7.3)^2 f(x)$$

$$= \frac{1}{80} \, [(2 - 7.3)^2 \, 8 + (3 - 7.3)^2 \, 15 + (4 - 7.3)^2 \, 20 + (5 - 7.3)^2 \, 10$$

$$+ (6 - 7.3)^2 \, 6 + (7 - 7.3)^2 \, 5 + (20 - 7.3)^2 \, 16]$$

$$= \frac{1}{80} \, (224.7 + 277.4 + 217.8 + 52.9 + 10.1 + 0.4 + 2580.6)$$

$$= \frac{1}{80} \, 3364 = 42.01$$

(c) What is the 90% confidence interval for the number of dollars spent per lunch for the population of all Chris's salespersons?

$$\mu = \bar{x} + z\frac{\bar{\sigma}}{\sqrt{N}}$$

$$\bar{\sigma} = \sqrt{\bar{\sigma^2}} = \sqrt{42.01} = \$6.5$$

$$\mu = \$7.3 \pm \$1.64 \left(\frac{6.5}{\sqrt{80}}\right)$$

$$= \$7.30 \pm \$1.20/\text{lunch}$$

or the average lunch costs between $6.10 and $8.50.

(d) What is the effect on the mean and standard deviation of adding these 16 heavy eaters?

Adding the 16 heavy eaters causes a substantial increase in the mean and the standard deviation.

7

Testing Hypotheses

In the model building process outlined in Chapter 1, once the data have been analyzed and the model is built, an hypothesis is made. This hypothesis is the best estimate or prediction of what will occur in a specific situation or an explanation of why certain situations produced certain results. The results generated mathematically by the application of the hypothesis are compared with the actual results or data achieved when a sample is taken, and a decision is then made concerning the validity of the hypothesis.

The hypothesis that we have made and shall assume is accurate is called the *null hypothesis*. It is represented by H_0. The *alternative hypothesis* H_i includes all possibilities excluded by the null hypothesis. If one is accurate, the other is false.

To test the null hypothesis, we collect sample data. If we are concerned, for example, with mean values, we see whether the sample mean $\bar{x}$ falls within *critical points*. Say we are checking the diameters of ball bearings that must be very close to the 0.5 cm value stated by the manufacturer in order to work. We hypothesize their diameters to have a mean value of 0.5 cm. We take a sample of 10 and measure them and find the mean for the sample. If we have established critical values of 0.49 and 0.51 cm and $\bar{N}$ falls between, we accept the hypothesis. If it does not, we reject it. How to establish critical points is discussed within this chapter.

If it turns out that we rejected a false hypothesis or accepted a true one, then the decision was correct. However, a problem arises when we reject as false a true hypothesis, or when we accept as true a false hypothesis. These errors are called type I and type II errors, respectively.

	True H_0	False H_0
Accept H_0	correct decision	type II error
Reject H_0	type I error	correct decision

The probability of a type I error is called α (alpha), and the probability of a type II error is called β (beta). The analysis of these two types must be considered separately.

TYPE I ERROR

If sample data are sufficiently close to the value of the null hypothesis, then the null hypothesis is accepted as being correct. How close is "sufficiently close," or, in other words, how close to H_0 must the critical points be, is a function of how confident one wishes to be of the results. The greater the closeness we require, the more likely it will be that we reject a value as false that is actually true (type I error). The less the closeness we require, the more likely it will be that we accept as true an hypothesis that is actually false (type II error).

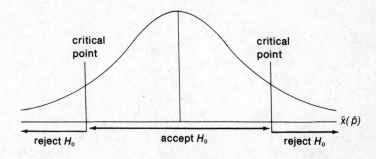

Figure 7.1

TWO-TAIL PROBLEMS

When, as above, the null hypothesis specifies one particular value, there are two rejection regions, one above the greater critical point, the other, below the lower critical point (Figure 7.1). This type of a problem is called *two-tailed*. The wording in one of these problems must indicate this feature; often it is one of the following: "test if the coin is fair," "exactly equals," "is the same as," "has not changed," "is different from."

ONE-TAIL PROBLEMS

In *two-tail* problems the concern is whether or not a sample value is approximately equal to the null hypothesis. In *one-tail* problems the concern is whether or not the sample mean is either at least or at most the value of the null hypothesis. A one-tail problem might concern whether a process is filling at least 12 oz in a bottle ($H_0 \geq 12$ oz), with the alternative being that less than 12 oz is put in the bottle ($H_i \leq 12$ oz).

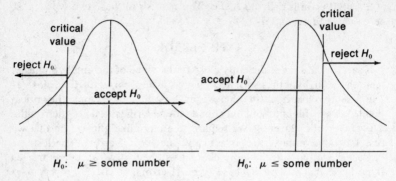

Figure 7.2

The key words for a one-tail test imply an inequality and the following are representative: "greater than," "less than," "at least," "more than," "less than," "minimum," "maximum." Figure 7.2 represents two cases of one-tailed tests.

TYPE II ERROR

If one accepts as true a false hypothesis, a type II error has occurred. The probability of a type II error, β, is a function of the false hypothesis, the sample size, and the degree of confidence specified in the situation.

Initially a region must be defined such that if the sample mean lies within it, then the null hypothesis will be accepted. The only region in which a false hypothesis could be accepted is within the "accept" region, for only there will an hypothesis be accepted. The accept region associated with the null hypothesis must be specified with the problem,

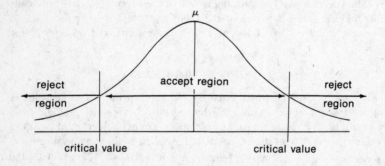

Figure 7.3

which may be accomplished by specifying the critical values or by providing a way of ascertaining the critical values (such as by specifying a desired level of confidence).

The problems at this stage appear the same as with type I (Figure 7.3). However, here we accept an hypothesis even though it turns out not to express the actual mean. However, the actual mean is not the one hypothesized: it is not within the critical values. The actual mean also has a normal curve surrounding it, so the probability of a type II error is the area under the curve of the actual mean between the critical values that were defined for the hypothesis.

For example, suppose one set a null hypothesis that $\mu = 50$ when actually $\mu = 60$. What is the probability that the false $\mu = 50$ would be accepted as true?

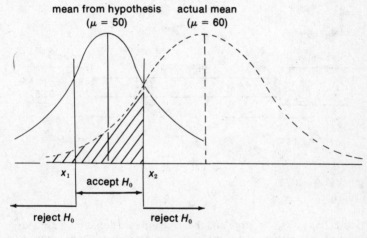

Figure 7.4

The problem is to determine the area under the dotted curve (actual mean distribution) between the critical values (x_1 and x_2) around the H_o. It is the shaded area in Figure 7.4. Once the critical values are obtained, the calculations merely involve solving for the area under a normal curve between two points.

The appropriate formulas are used to generate a z number for each of the x_1 and x_2 points and the areas are looked up in the table (Appendix I). The difference in the areas is β.

Example 7.1

A survey of income for union truck drivers yielded the following results in dollars per hour: 2, 4, 7, 1, 3, 3, 4, 2, 8, 6. A labor relations expert hypothesized that exactly 60% of the drivers make less than $5/h. Evaluate her estimate with 90% confidence (do this in percentages). Note that the sample has 7 people, or 70%, earning less than $5/h.

$$H_0: \quad \mu = p = 0.6$$
$$H_i: \quad p \neq 0.6$$

We want 90% of the area to lie between the two critical points, so from the mean to either critical point should contain ½ of 0.90, or 0.45, of the area. Looking up a 0.45 area in the table and reading a z number yields $z = 1.64$.

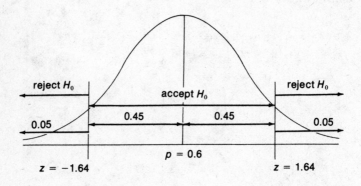

Figure 7.5

Figure 7.5 shows the curve and the decision rule derived; i.e., if z from the sample is between -1.64 and 1.64, then we accept the hypothesis as being correct, but if z from the sample is less than -1.64 or greater than 1.64, we reject the H_0 and accept H_i. We use the equation for p from the "Proportion" section of Chapter 6, which we solve for z:

$$\bar{z} = \frac{\bar{p} - p}{\sqrt{\dfrac{p(1 - p)}{N}}} = \frac{0.7 - 0.6}{\sqrt{\dfrac{0.6(0.4)}{10}}} = 0.65$$

We accept H_0. Here $\bar{p}$ is the proportion for the sample.

The next example is very similar in format, but we have changed the values of several of the variables to illustrate the effect of such changes. Note that the basic format and the formula for z of the sample are the same. Varying the confidence limits changes the critical points, and changing the hypothesis moves the center point around which the critical points are measured and changes the input in the formula for z of the sample. The conclusion rests on the sample values and their relationship to the standards, which is as might be anticipated, for these are the only inputs from the actual process being evaluated. The other inputs are statistical standards used to evaluate the sample quantities.

Example 7.2

A survey of income for nonunion truck drivers yielded the following results in dollars per hour: 8, 2, 4, 7, 4, 6, 7, 10. A labor relations expert hypothesized that exactly 50% of the drivers make less than $5/h. Evaluate her estimate with 95% confidence (use percentages). Note that 3 people actually made less than $5/h.

$$H_0: \quad \mu = p = 0.50$$
$$H_i: \quad \mu \neq 0.50$$

We want 95% of the area to lie between the two critical points, so from the mean to either critical point should contain ½ of 0.95, or 0.475 in area. Looking up an 0.475 area in the table and reading a z number yields $z = 1.96$.

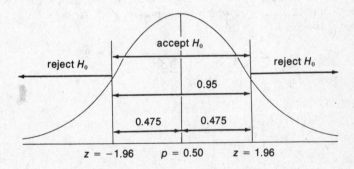

Figure 7.6

Figure 7.6 shows the curve and the decision rule derived; i.e., if z from the sample is between -1.96 and 1.96, then we accept the hypothesis as being correct, but if z from the sample is less than -1.96 or greater than 1.96, we reject it.

The labor relations expert then hypothesized that at least 50% make less than \$5/h. Should we accept or reject this hypothesis?

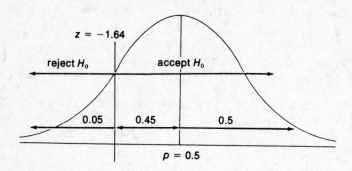

Figure 7.7

The mean value is considered as if the problem were an equality, and the critical point is to the left because if the sample value is very close to the mean but a little less, we would not wish to reject the hypothesis. The formula for the sample z, which will represent as $\bar{z}$, is the same as in previous problems.

$$\bar{z} = \frac{\bar{p} - p}{\sqrt{\dfrac{p(1 - p)}{N}}} = \frac{0.375 - 0.5}{\sqrt{\dfrac{0.5(0.5)}{8}}} = -0.7$$

We accept H_0.

Example 7.3

Again we consider the same survey. The labor relations expert hypothesized that at *most* 25% of the drivers made more than \$6/h. Evaluate her estimate with 85% confidence (use percentages).

$$H_0: \quad p \leq 0.25$$
$$H_i: \quad p > 0.25$$

We want 85% of the area to lie in the accept region and 15% in the reject region. This is a one-tail test because we want to prove an inequality, i.e., that at most 25% did something, so if 0% did it, we would still accept the hypothesis.

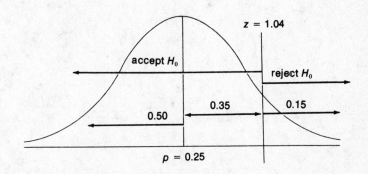

Figure 7.8

The critical point ($z = 1.04$) was found by looking up 0.35 in the table and reading out a z value (Figure 7.8). The formula for $\bar{z}$ (of the sample) is still the same.

$$\bar{z} = \frac{\bar{p} - p}{\sqrt{\dfrac{p(1-p)}{N}}} = \frac{0.50 - 0.25}{\sqrt{\dfrac{0.25(0.75)}{8}}} = 1.63$$

We reject H_0.

Example 7.4

You hypothesize that the average statistics quiz grade is 75%, with a variance of 16%. With 99% confidence and based on sample values of a 76.5% average for 64 students, evaluate your hypothesis against the situation that the average grade has changed from 75.

$$H_0: \quad \mu = 75\%$$
$$H_i: \quad \mu \neq 75\%$$

You want to establish critical points such that 99% of the area is between them (Figure 7.9). From the mean to a critical point there should be ½ of 0.99, or 0.495. Looking up 0.495 in the center of the table and reading back out to the margins yields a z number of 2.58.

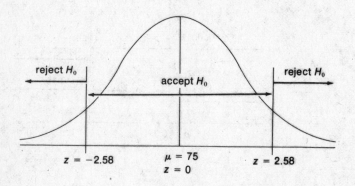

Figure 7.9

$$z = \frac{x - \mu}{\frac{\sigma}{\sqrt{N}}} = \frac{76.5 - 75}{\frac{4}{\sqrt{64}}} = 3.00$$

Since a z value of 3.00 falls within the reject region, we reject H_0 and accept H_i. Note that since the formula calls for σ (standard deviation), we had to take the square root of the variance to use the formula.

Example 7.5

You hypothesize that the average grade on a statistics quiz is 75%, with a variance of 16%. Based on this hypothesis you establish critical points of 78% and 72%. If a sample value is between these two points, you will accept your hypothesis. If the mean is actually 79%, what is the probability that a mean from a sample of 36 will fall within your critical points of 78% and 72%? This probability of accepting a false hypothesis is the probability of a type II error, or β.

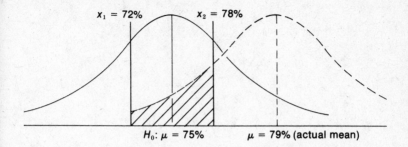

Figure 7.10

The problem has evolved to one of determining the area under the curve between the two critical points, which involves calculating a z number, looking up an area in the table for the z number, and then possibly manipulating these areas (Figure 7.10).

$$z_1 = \frac{x_1 - \mu}{\dfrac{\sigma}{\sqrt{N}}} = \frac{72 - 79}{\dfrac{4}{\sqrt{36}}} = -10.5$$

$$z_2 = \frac{x_2 - \mu}{\dfrac{\sigma}{\sqrt{N}}} = \frac{78 - 79}{\dfrac{4}{\sqrt{36}}} = -1.5$$

$$\beta = A_{z = -10.5} - A_{z = -1.5} = 0.5 - 0.4332 = 0.0668$$

There is approximately a 7% chance of accepting the H_0 when the mean is actually 79%. Note that the probability of accepting a false hypothesis depends on the value of the actual mean. See next example for demonstration.

Example 7.6

Consider the previous example but let the actual mean be 77% (Figure 7.11).

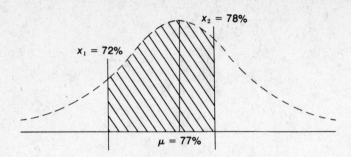

Figure 7.11

$$z_1 = \frac{x_1 - \mu}{\dfrac{\sigma}{\sqrt{N}}} = \frac{72 - 77}{\dfrac{4}{\sqrt{36}}} = -7.5$$

$$z_2 = \frac{x_2 - \mu}{\dfrac{\sigma}{\sqrt{N}}} = \frac{78 - 77}{\dfrac{4}{\sqrt{36}}} = 1.5$$

$$\beta = A_{z = 1.5} + A_{z = -7.5} = 0.4332 + 0.5 = 0.9332$$

Compare this example with 7.5 and note that as the null hypothesis is chosen closer to the actual mean, the probability of accepting the null hypothesis as the true mean is increased.

PROBLEMS

Problem 7.1

A student on a true–false test got out of 100 questions 85 correct. The teacher feels that the student should be doing exactly 96% work. With 90% confidence, evaluate the teacher's feeling.

$$H_0: \quad \mu = p = 0.96$$
$$H_i: \quad p \neq 0.96$$

For 90% confidence, 90% of the area should lie between the two critical points, so from the mean to either critical point should be ½ of 90%, or 0.45 in area (Figure 7.12). Looking up 0.45 in the z table yields $z = 1.65$.

The z for the sample is

$$\bar{z} = \frac{\bar{p} - p}{\sqrt{\dfrac{p(1 - p)}{N}}}$$

$$= \frac{0.85 - 0.96}{\sqrt{\dfrac{0.96(1 - 0.96)}{100}}} = \frac{0.85 - 0.96}{\sqrt{\dfrac{0.96(0.04)}{100}}} = -5.5$$

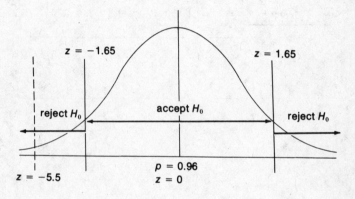

Figure 7.12

Reject H_0 because $\bar{z}$ lies within the reject region.

Problem 7.2

Consider the same student (85 right out of 100) and a second teacher's opinion that the student should get at least 88 correct. Evaluate this opinion at the 95% confidence level.

$$H_0: \quad \mu = 12 \text{ oz}$$
$$H_i: \quad \mu \neq 12 \text{ oz}$$

95% of the area should lie within the accept region; so 5% must lie in the reject region (Figure 7.13). We are trying to determine whether a minimum standard is met, so there is only one reject region. Thus, this is a one-tail test. For the critical point, look up an area of 0.45, which leaves 0.05 in the tail. This gives a z value of 1.64.

$$\bar{z} = \frac{\bar{p} - p}{\sqrt{\dfrac{p(1 - p)}{N}}} = \frac{0.85 - 0.88}{\sqrt{\dfrac{0.88(1 - 0.88)}{100}}}$$

$$= -1.0$$

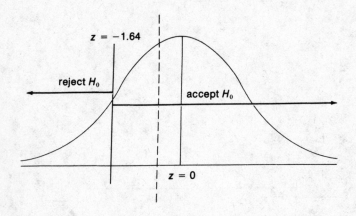

Figure 7.13

Since $\bar{z} = -1.0$, we accept H_0.

Problem 7.3

Again consider the same student (85 correct out of 100) and a third teacher's claim that the student should get at most 75 correct. Evaluate this claim with 95% confidence.

$$H_0: \quad p \leq 0.75$$
$$H_i: \quad p > 0.75$$

95% of the area should lie in the accept region and 5% in the reject region (Figure 7.14). This is a one-tail test because you want to prove an inequality. Look up an area of 0.45 in the z table. The z value is 1.65.

$$\bar{z} = \frac{\bar{p} - p}{\sqrt{\dfrac{p(1-p)}{N}}} = \frac{0.85 - 0.75}{\sqrt{\dfrac{0.75(1-0.75)}{100}}} = \frac{0.10}{0.043}$$

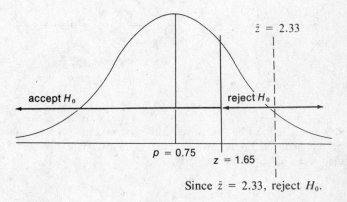

$\bar{z} = 2.33$

accept H_0

reject H_0

$p = 0.75$ $z = 1.65$

Since $\bar{z} = 2.33$, reject H_0.

Figure 7.14

Problem 7.4

You think that the average weight of a package received in your mail room is 12 oz, with a variance of 9 oz. With 95% confidence, based on a sample mean of 10, a mode of 6, a median of 9, and a sample size of 25, evaluate your estimate.

$$H_0: \quad \mu = 12 \text{ oz}$$
$$H_i: \quad \mu \neq 12 \text{ oz}$$

$$\bar{z} = \frac{\bar{x} - \mu}{\dfrac{\sigma}{\sqrt{N}}} = \frac{10 - 12}{\dfrac{3}{\sqrt{25}}} = \frac{-2}{\dfrac{3}{5}} = -3.33$$

$$\sigma = \sqrt{\sigma^2} = \sqrt{9} = 3$$

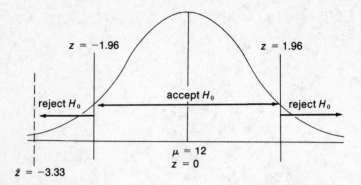

Since $\bar{z} = -3.33$, reject H_0.

Figure 7.15

Note that one must be able to select the necessary items for a solution to be reached and reject the unnecessary items. Here the mode and median were not necessary (Figure 7.15).

Problem 7.5

You think that the average weight on packages you receive is 12 oz, with a variance of 9 oz. Based on this, you establish critical points of 6 oz and 18 oz, so that if a sample value is between these two points you will accept your estimate as being correct. If the mean is actually 16 oz, what is the probability that a sample at 25 oz will fall within your critical points of 6 oz and 18 oz, and that you will accept your hypothesis that the average is 12 oz when the actual average is 16 oz. This probability of accepting a false hypothesis is the probability of a type II error, or β.

The solid curve is drawn from the H_0, and the dashed-line curve reflects the actual situation (Figure 7.16).

The question is, what is the area under the dashed curve between the values established as critical for the hypothesis? The problem has evolved to one of determining the area under the curve between two points, which always involves calculating a z number, looking up an area in the table for the z number, and then possibly manipulating these areas.

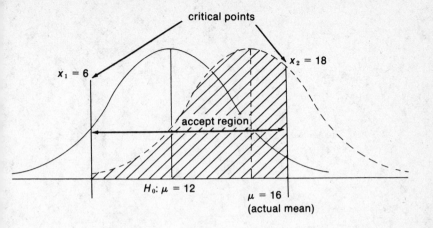

critical points

$x_1 = 6$

$x_2 = 18$

accept region

$H_0: \mu = 12$

$\mu = 16$
(actual mean)

Figure 7.16

$$z_1 = \frac{x_1 - \mu}{\frac{\sigma}{\sqrt{N}}} = \frac{6 - 16}{\frac{3}{\sqrt{25}}} = \frac{-10}{\frac{3}{5}} = -16.67$$

$$z_2 = \frac{x_2 - \mu}{\frac{\sigma}{\sqrt{N}}} = \frac{18 - 16}{\frac{3}{\sqrt{25}}} = \frac{2}{\frac{3}{5}} = 3.3$$

$$\beta = A_{z = -16.67} + A_{z = 3.3}$$

$$A_{z = -16.67} = 0.5$$

Since if a z number is too large for the table, one should take the highest value from the table

$$A_{z = 3.3} = 0.4988$$

$$\beta = 0.5 + 0.4988 = 0.9988$$

Problem 7.6

Consider the previous example, but let the actual mean be 20. The critical points and sample data are the same, as is your hypothesis.

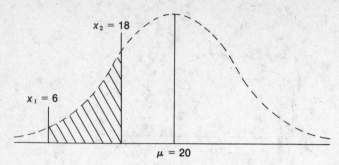

Figure 7.17

(See Figure 7.17.)

$$z_1 = \frac{x_1 - \mu}{\dfrac{\sigma}{\sqrt{N}}} = \frac{6 - 20}{\dfrac{3}{\sqrt{25}}} = -23.3$$

$$z_2 = \frac{z_2 - \mu}{\dfrac{\sigma}{\sqrt{N}}} = \frac{18 - 20}{\dfrac{3}{\sqrt{25}}} = -3.33$$

$$\beta = A_{z = -23.3} - A_{z = -3.33} = 0.5 - 0.4988 = 0.0012$$

8

Small Samples—t Distribution

In small samples the t distribution is more appropriate than the z statistic to use for problem solving. A rule of thumb is that the t distribution should be used when the sample size N is less than 30. The operations that can be performed by using the t distribution and the t statistic are very similar to those that can be performed by using the z statistic. Indeed, the larger the samples, the closer are the values of the t statistic to the values of the z statistic. In model building, the t distribution is used when the data have approximately a normal distribution.

In addition, the format of the problems is similar, and there is a table (Appendix II) that gives certain specified values of t.

There are two differences between t and z problems: The formulas are slightly different and the tables with the values are arranged differently. The formulas for the t distribution are these

$$t = \sqrt{N-1} \left(\frac{x - \mu}{\tilde{\sigma}} \right) \tag{1}$$

or solving for μ,

$$\mu = x \pm t \frac{\tilde{\sigma}}{\sqrt{N-1}} \tag{2}$$

Formula (1) is used for hypothesis testing and (2) is used for making inferences about the population based on data about the sample. When testing an hypothesis, as with z problems, an alternative hypothesis (H_i) must be established that includes all the outcomes not included in H_0.

USE OF THE t TABLE

The table for t gives, for certain confidence levels, the critical values of the t statistic. The values in the body of the table are the critical t values, and the confidence levels are given in the margins at the top of the table; note that this is a reversal of the z table, where the z levels are given within the table rather than in the margins. For example, using the t table for a 95% confidence level and 6 degrees of freedom yields a critical t value of 2.45. Degrees of freedom are equal to the sample size minus 1.

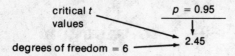

Remember that the total of the confidence level plus the level of significance equals 1. Note that only the certain specified confidence levels that are listed in the margin of the table can be used directly. To use the table for one-tail values we must make the calculation $1 - 1/2$ level of significance. This gives the appropriate column. Where it crosses the appropriate row the t value is located. The appropriate row is located by calculating the degree of freedom, which equals the sample size minus 1. Degrees of freedom may be thought of in the following manner: If the total of the items in the sample is known, how many items are we free to choose or vary the values for? Answer: We are "free" to vary the values for all but the last one, i.e., there are $(N - 1)$ degrees of freedom.

The sample size is the controlling factor in deciding whether to use a z or t statistic; if the sample is less than 30, then use the t statistic.

Example 8.1

You hypothesize that the average statistics quiz grade is 75, with a variance of 16 (Figure 8.1). With 99% confidence, based on a sample mean value of 76.5 for 64 students, evaluate your hypothesis. (Note that between the null and alternative hypotheses every possible outcome is included.)

The sample size is 64, which is greater than 30, so use the z statistic.

$$H_0: \quad \mu = 75$$
$$H_i: \quad \mu \neq 75$$

$$z = \frac{x - \mu}{\dfrac{\sigma}{\sqrt{N}}} = \frac{76.5 - 75}{\dfrac{4}{\sqrt{64}}} = \frac{1.5}{0.5} = 3.0$$

Since z lies outside the accept region, we reject H_0.

Example 8.2

Again hypothesize that the average statistics quiz grade is 75, with a variance of 16. With 95% confidence, based on a sample average of 74 for

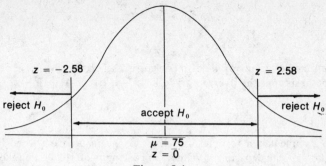

Figure 8.1

10 students, evaluate your hypothesis against the alternative that the average grade has changed from 75.

The sample size is 10, which is less than 30, so use the *t* statistic.

$$H_0: \quad \mu = 75$$
$$H_i: \quad \mu \neq 75$$

To find the critical *t* statistic values, one must know the degrees of freedom (sample size minus 1) and the percent confidence desired. With 95% confidence desired, the critical *t* value for a sample of, e.g., 10 degrees of freedom, is 2.26 (Figure 8.2).

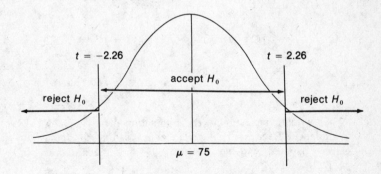

Figure 8.2

The decision rule is that if the *t* calculated from the sample, represented as $\bar{t}$, is between −2.26 and 2.26, then accept the hypothesis. If the *t* from the sample is less than −2.26 or greater than 2.26, then reject the hypothesis.

$$t = \sqrt{N - 1}\, \frac{x - \mu}{\bar{\sigma}} = \sqrt{10 - 1}\, \frac{74 - 75}{4}$$

$$= -0.75$$

The sample value for $\bar{t}$ lies between -2.26 and 2.26, so accept the hypothesis that the average grade is 75.

Example 8.3

A sample taken to determine the average insurance carried by the workers of one department yielded, in base units of $1000, the following: 2, 6, 11, 9, 5, 9.

(a) What would you estimate the exact amount of insurance carried by all workers to be? The best point estimate for the population mean μ is the sample mean $\bar{x}$.

$$\bar{x} = \sum_{i=1}^{6} x_i \, \frac{f(x_i)}{N} = \frac{1}{6} (2 + 6 + 11 + 9 + 5 + 9) = 7$$

(b) What is the standard deviation(s) of the sample?

standard deviation $= \sqrt{\text{variance}}$

$$\sigma = \sqrt{\sigma^2}$$

$$\bar{\sigma}^2 = \sum_{i=1}^{6} (x_i - \bar{x})^2 \, \frac{f(x_i)}{N} = \frac{1}{N} \sum_{i=1}^{6} (x_i - \bar{x})^2 f(x_i)$$

$$= \frac{1}{6} [(2 - 7)^2 + (6 - 7)^2 + (11 - 7)^2$$

$$+ (9 - 7)^2 + (5 - 7)^2 + (9 - 7)^2] = 9$$

$$\bar{\sigma} = 3$$

(c) What would you estimate the population mean to be with 95% confidence? Because the sample size is less than 30, the t statistic should be used, with $(N - 1)$, or $(6 -)$, degrees of freedom.

$$\mu = \bar{x} \pm t \, \frac{\bar{\sigma}}{\sqrt{N - 1}} = 7 \pm 2.57 \, \frac{3}{\sqrt{6 - 1}} = 7 \pm 3.44$$

This means that based on the sample and the given confidence level that the mean for the population should lie between 3.56 and 10.44,

or $3.56 \le \mu \le 10.44$

(d) To illustrate the effect of the sample size, hold the other factors the same and increase the sample size to 26, which will change the *t* value.

$$\mu = \bar{x} \pm t \, \frac{\bar{\sigma}}{\sqrt{N-1}} = 7 \pm 2.06 \, \frac{3}{\sqrt{26-1}}$$

$$= 7 \pm 1.24$$

or $5.76 \leq \mu \leq 8.24$

The increased sample size results in a much narrower range for the population mean.

(e) To illustrate the effect of the standard deviation, hold all the parts of the previous example constant except for the standard deviation, which change to 20.

$$\mu = \bar{x} \pm t \, \frac{\bar{\sigma}}{\sqrt{N-1}} = 7 \pm 2.06 \, \frac{20}{\sqrt{26-1}}$$

$$= 7 \pm 8.24$$

or $-1.24 \leq \mu \leq 15.24$

The increased dispersion among the sample values, and hence increased standard deviation, increases the range in which one would expect to find the population mean.

PROBLEMS

Problem 8.1

The Department of Health, Education and Welfare (HEW) estimates that the average monthly income for people in a given area is $2500, with a standard deviation of $100. With 95% confidence, based on a sample mean value of $3000 income for 17 people sampled, evaluate HEW's hypothesis against Senator Watchem's claim that the average has changed.

The sample is less than 30, so use the *t* statistic.

$$H_0: \quad \mu = 2500$$
$$H_i: \quad \mu \neq 2500$$

Find the critical value for *t*.

$$\text{degrees of freedom} = N - 1 = 17 - 1 = 16$$
$$\% \text{ confidence} = 95\%$$
$$\text{critical value of } t = 2.12$$

(A two-tail test is appropriate for the H_i because the issue is whether "the average has changed"; i.e., whether or not μ equals a particular value. (See Figure 8.3.)

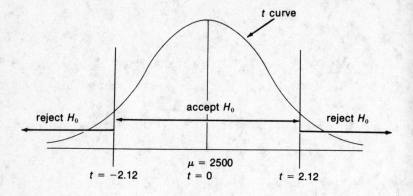

Figure 8.3

Now the t sample $\bar{t}$ must be calculated. Depending on where its value is located with respect to the critical values of t, we either accept or reject the H_0.

$$\bar{t} = \sqrt{N-1}\left(\frac{\bar{x} - \mu}{\bar{\sigma}}\right)$$

$$= \sqrt{17-1}\left(\frac{\$3000 - \$2500}{\$100}\right)$$

$$= 20$$

The sample value for $\bar{t}$ is outside the accept region for t (between -2.12 and $+2.12$), so we reject the H_0 and accept the H_i (that the income level is *not* $2500).

Problem 8.2

Congresswoman Worrell views the same data as the Senator did in the previous example, but she feels that the income level has *increased*. Again, HEW feels that the average monthly income is $2500 or less, with a standard deviation of $100. The sample mean is $3000 for 17 people. Evaluate HEW's claim in light of the Congresswoman's ideas, with 90% confidence.

$$H_0: \quad \mu \leq \$2500$$
$$H_i: \quad \mu > \$2500$$

Notice that the problem's wording ("the income level has increased") leads to an inequality test and hence a one-tail test. Find the critical value for *t*.

$$\text{degrees of freedom} = N - 1 = 17 - 1 = 16$$
$$\% \text{ confidence} = 90\%$$
$$\text{critical value of } t = t[1 - \tfrac{1}{4}(0.1)] = t(0.95) = 2.12$$

Note that the critical value is the same as for the above example, but the percent confidence is different, because here there is a one-tail test (Figure 8.4).

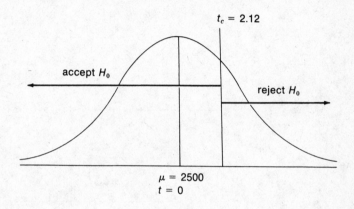

Figure 8.4

Now the $\bar{t}$ must be calculated. Depending on where its value is located with respect to the critical values of *t*, we either accept or reject the H_0.

$$\bar{t} = \sqrt{N - 1}\left(\frac{\bar{x} - \mu}{\bar{\sigma}}\right) = \sqrt{17 - 1}\left(\frac{3000 - 2500}{100}\right) = 20$$

The sample value for $\bar{t}$ is outside the accept region for *t* (less than 2.12), so we reject the H_0 and accept the H_i, which is that the income level has *increased* from $2500.

Problem 8.3

Administrator Grant, representing HEW, views the data in the above examples, and changes his hypothesis to be that the average monthly income is *at least* $2500, with a standard deviation of $100. The mean for the sample of 17 people is still $3000. Evaluate HEW's claim with 90% confidence.

$$H_0: \quad \mu \geq \$2500$$
$$H_i: \quad \mu < \$2500$$

Note that the problem's wording "is at least" leads to an inequality test and hence a one-tail test (Figure 8.5). Find the critical value for t.

$$\text{degrees of freedom} = N - 1 = 17 - 1 = 16$$
$$\% \text{ confidence} = 90\%$$
$$\text{critical value of } t = t[1 - \tfrac{1}{2}(0.1)] = t(0.95) = 2.12$$

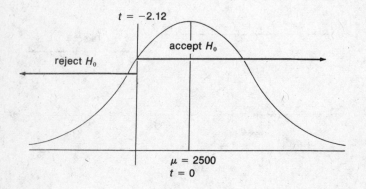

Figure 8.5

Note that because of the "greater than" H_0, the critical value is on the left side of the mean. Hence that critical t has a negative sign. This is always the procedure when the H_0 states that μ is greater than some specified value.

Now the $\bar{t}$ must be calculated. Depending on where its value is located with respect to the critical values of t, we either accept or reject H_0.

$$\bar{t} = \sqrt{N - 1}\left(\frac{\bar{x} - \mu}{\bar{\sigma}}\right) = \sqrt{17 - 1}\left(\frac{3000 - 2500}{100}\right)$$
$$= 20$$

The value for $\bar{t}$ is *within* the accept region for H_0, so we accept H_0, that the average income is greater than $2500.

Problem 8.4

Compare the previous three examples and note that the data are the same for all three but the hypotheses are different. What effect does this have?

The examples demonstrate these effects. First, changing the hypotheses, as in the above examples, does not have any effect on $\bar{t}$ because the inequalities are converted to equalities when we do the calculating. The difference in the problems lies in how t from the table is selected and utilized to establish a "test."

The t from the table is selected somewhat differently depending on whether a one- or a two-tail test is to be used. Remember that a one-tail test indicates an inequality.

The accept and reject regions are considerably different in a one-tail test from the way they are in a two-tail test. The difference affects whether or not a particular hypothesis is accepted but does not affect the final result. In the last two examples, the final result is that the mean income is greater than $2500, which result is achieved under two different hypotheses: in one example the H_0 was accepted and in the other example the H_0 was rejected. Thus, the final result is the same regardless of which of the one-tailed tests was selected.

The result from the two-tailed test was a little different, because all one can prove with this test is whether or not something is equal to the H_0. One cannot prove whether or not the sample values are actually larger or smaller than the standard. The one-tail test enables one to establish larger or smaller results. The same is true for z statistic problems, only the samples are larger than 30.

Problem 8.5

Anne, an office manager, takes a sample to determine the average age of secretaries in an office and obtained the results 23, 47, 51, 30, 40, 36, 49, 33, 29, 52, 36. What should Anne's estimate be for the average age for all secretaries in the building based on this sample.

The appropriate statistic to use is the t statistic because the sample size ($N = 11$) is less than 30. With the same problem with an $N > 30$ one would use the z statistic.

The point estimate for the population is the sample mean, which is the same regardless of the statistic used.

$$\bar{x} = \frac{1}{N} \sum_{i=1}^{11} x_i f(x_i)$$

$$= \frac{1}{11} [23(1) + 47(1) + 51(1) + 30(1) + 40(1) + 36(2)$$

$$+ 49(1) + 33(1) + 29(1) + 52(1)]$$

$$= \frac{1}{11} (426) = 38.7 \text{ years old}$$

The 95% confidence estimate for the mean involves a statistic, in this case, the t statistic. So it is necessary to calculate or look up in a table the following: the t values of the critical points, $\bar{\sigma}$, $N - 1$, and $\bar{x}$ (which has already been calculated).

$$N - 1 = 11 - 1 = 10$$

critical value of t for $N - 1$, or 10, degrees of freedom and 95% confidence

$$t = 2.23$$

$$\bar{\sigma}^2 = \text{ sample standard deviation}$$

(which requires first calculating the variance)

$$\bar{\sigma}^2 = \frac{1}{N} \sum_{i=1}^{N} (x_i - \bar{x})^2 f(x_i)$$

$$= \frac{1}{11} [(23 - 38.7)^2\, 1 + (47 - 38.7)^2\, 1 + (51 - 38.7)^2\, 1$$

$$+ (30 - 38.7)^2\, 1 + (40 - 38.7)^2\, 1 + (36 - 38.7)^2\, 2$$

$$+ (49 - 38.7)^2\, 1 + (33 - 38.7)^2\, 1 + (29 - 38.7)^2\, 1$$

$$+ (52 - 38.7)^2\, 1]$$

$$= \frac{1}{11} (246.5 + 68.9 + 151.3 + 75.7 + 1.7 + 14.6 + 10.3$$

$$+ 32.5 + 94.1 + 176.9)$$

$$= \frac{1}{11} (872.5) = 79.3$$

$\bar{\sigma} = \sqrt{79.3} = 8.9 \text{ years old}$

Back to the formula for the estimate for the mean:

$$\mu = \bar{x} \pm t \left(\frac{\bar{\sigma}}{N - 1} \right) = 38.7 \pm 2.23 \frac{8.9}{\sqrt{11 = 1}}$$

$$= 38.7 \pm 6.3$$

Or the average age for secretaries in the office is between 32.4 and 45.0 years old.

9

Difference Between Means

Here we will discuss the way in which two samples or mean values of two samples can be compared. These might be samples from two different populations or samples of the same population taken at different times. Two different samples taken at random from the same population might be expected to yield different mean values because some change might have taken place in the population during the interval between the samplings or because one of the samples does not adequately represent the population it was taken from.

The technique used in comparing samples is to hypothesize that there is no difference in the means of the two populations (or one population measured at two different times). The appropriate standard values for the t statistic, based on the percent confidence desired and the degrees of freedom (to be discussed later), are obtained from the t table, and a test is established using critical values of the t statistic (see Figure 9.1).

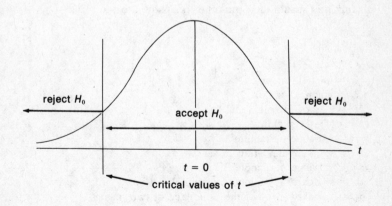

Figure 9.1

The center point of the curve represents the value that the t statistic would have if the two means were the same. A t value based on the sample data is calculated, using the formula presented below, and this calculated $\bar{t}$ value is compared to the test, or critical values, of t. If $\bar{t}$ is sufficiently close to $t = 0$ (or the center), i.e., is within the critical values, then the hypothesis is accepted, but if $\bar{t}$ is not sufficiently close to $t = 0$, then the hypothesis is rejected. How close is "sufficiently close" is determined by the size of the samples (as reflected by the degrees of freedom) and the desired confidence in the result.

The test or critical values of the t statistic are found in the table, in Appendix II, under the specified percent confidence.

The degrees of freedom in this t test equal the size of sample 1 minus 1 plus the size of sample 2 minus 1, or $N_1 + N_2 - 2$.

THE t STATISTIC

The formula for $\bar{t}$, from the samples, is

$$\bar{t} = \frac{\bar{x}_1 - \bar{x}_2}{\sqrt{\dfrac{\sigma_1{}^2}{N_1} + \dfrac{\sigma_2{}^2}{N_2}}}$$

Sometimes the standard error of the difference, σ_D, is desired, which is the denominator of the t statistic:

$$\sigma_D = \sqrt{\frac{\sigma_1{}^2}{N_1} + \frac{\sigma_2{}^2}{N_2}}$$

Making this substitution puts the t statistic into the form:

$$\bar{t} = \frac{\bar{x}_1 - \bar{x}_2}{\sigma_D}$$

where

$\bar{x}_1$ = mean of sample 1
$\bar{x}_2$ = mean of sample 2
N_1 = size of sample 1
N_2 = size of sample 2
$\sigma_1{}^2$ = variance of sample 1
σ_1 = standard deviation of sample 1
$\sigma_2{}^2$ = variance of sample 2
σ_2 = standard deviation of sample 2
σ_D = standard error of the difference of two means
$\bar{t}$ = t statistic calculated for the samples

Example 9.1

Samples are taken from a machine at two different times. At time 1, 50 items are selected, and their mean diameter is 0.50 in., with a standard deviation of 0.10 in. The 60 parts taken at time 2 have a mean diameter of 0.52 in., with a standard deviation of 0.15 in. The manager wishes to know with 95% confidence whether the machine's output has changed significantly.

Start with the hypothesis that there has been no change, the alternative being that a change has occurred.

H_0: The mean of the population implied by sample 1 equals the mean implied by sample 2 ($\mu_1 = \mu_2$)

H_i: $\mu_1 = \mu_2$

Then calculate the σ_D and the $\bar{t}$.

$$\sigma_D = \sqrt{\frac{\sigma_1{}^2}{N_1} + \frac{\sigma_2{}^2}{N_2}} = \sqrt{\frac{(0.10)^2}{50} + \frac{(0.15)^2}{60}}$$

$$= \sqrt{\frac{0.01}{50} + \frac{0.023}{60}} = \sqrt{0.0006} = 0.024$$

$$\bar{t} = \frac{\bar{x}_1 - \bar{x}_2}{\sigma_D} = \frac{0.50 - 0.52}{0.024} = \frac{-0.02}{0.024} = -0.83$$

Then look up in the table the critical values of t based on degrees of freedom and percent confidence. Degrees of freedom = $N_1 + N_2 - 2 = 50 + 60 - 2 = 108$. If you look down the margin of the table under degrees of freedom, note that there is one line after the line for 25 degrees of freedom. The appropriate numbers, depending on the percent of confidence desired (which dictates the column) are used for any problem where the degrees of freedom are greater than 25. In this example, for 95% confidence, the critical t values are ± 1.96. The test now looks as represented in Figure 9.2.

The decision rule is, If the $\bar{t}$ is greater than -1.96 or less than $+1.96$, then accept H_0 as being correct. If the $\bar{t}$ is less than -1.96 or greater than 1.96, then reject the hypothesis and accept the H_i.

The $\bar{t}$ value of -0.83 is within the accept region, so accept the hypothesis that there has been no change and the samples were drawn from populations with similar means.

Example 9.2

An engineer wishes to know with 95% confidence whether or not there is a significant difference between the time it takes for a Buick to stop and

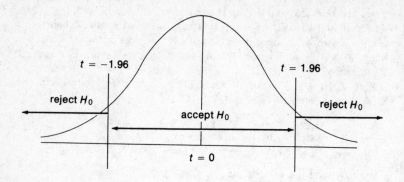

Figure 9.2

the time required for an Oldsmobile to stop. A sample of 6 Buicks were
tested, with the mean time to stop being 15 sec, with a standard deviation
of 6 sec. A sample of 5 Oldsmobiles yielded the mean time of 7 sec, with a
standard deviation of 4 sec.

To start, one always hypothesizes that there was *not* a difference.

$$H_0: \quad \mu_{\text{Buicks}} = \mu_{\text{Olds}}$$
$$H_i: \quad \mu_\text{B} \neq \mu_\text{O}$$

Then calculate σ_D and $\bar{t}$:

$$\sigma_D = \sqrt{\frac{{\sigma_1}^2}{N_1} + \frac{{\sigma_2}^2}{N_2}} = \sqrt{\frac{6^2}{6} + \frac{4^2}{5}} = \sqrt{\frac{36}{6} + \frac{16}{5}}$$

$$= \sqrt{6 + 3.2} = 3.03$$

$$\bar{t} = \frac{\bar{x}_1 - .\bar{x}_2}{\sigma_D} = \frac{15 - 7}{3.03} = 2.64$$

Find the critical values of t:

degrees of freedom $= N_1 + N_2 - 2 = 6 + 5 - 2 = 9$

t for 95% confidence and 9 degrees of freedom $= 2.26$

The decision rule is, if $\bar{t}$ is greater than -2.26 or less than 2.26, then
accept the hypothesis. If $\bar{t}$ is less than -2.26 or greater than 2.26, then
reject the H_0 and accept the H_i.

The $\bar{t}$ is 2.64: reject the H_0 that there is no significant difference between
the mean time that it takes a Buick and an Olds to stop.

Example 9.3

To illustrate the effect of the sample sizes, take the same data in the previous problem but increase the sample sizes. Sample 60 Buicks and 50 Oldsmobiles.

The hypothesis is the same:

$$H_0: \quad \mu_B = \mu_O$$
$$H_i: \quad \mu_B \neq \mu_O$$

Then calculate σ_D and $\bar{t}$:

$$\sigma_D = \sqrt{\frac{\sigma_1^2}{N_1} + \frac{\sigma_2^2}{N_2}} = \sqrt{\frac{(6)^2}{60} + \frac{(4)^2}{50}}$$

$$= \sqrt{0.6 + 3.2} = 1.95$$

$$\bar{t} = \frac{15 - 7}{1.95} = 4.10$$

Not only has the $\bar{t}$ changed, but the critical values for the t statistic are also different because of the change in the degrees of freedom.

$$\text{degrees of freedom} = N_1 + N_2 - 2 = 60 + 50 - 2 = 108$$

The critical t values are ± 1.96. Notice that increasing the sample size makes the test more precise, i.e., reduces the range for acceptable sample values.

That $\bar{t}$ is 4.10 still indicates that one should reject the hypothesis.

Example 9.4

Here we illustrate the effect of varying the standard deviations of the samples. Intuitively one feels that as the standard deviation increases, a greater dispersion between the values within each sample arises, so that it becomes more difficult to identify precisely any sample set and exclude other sets from the same set. If this is true, it will be reflected in a decrease in the value of $\bar{t}$.

Changing the standard deviations but keeping the other variables constant leads to the following data:

Buicks	Oldsmobiles
$\bar{x}_B = 15$	$\bar{x}_O = 7$
$\sigma_B = 25$	$\sigma_O = 40$
$N_B = 6$	$N_O = 5$

Again hypothesizing no difference, we have

$$H_0: \quad \mu_O = \mu_B$$
$$H_i: \quad \mu_O = \mu_B$$

Then calculate σ_D and $\bar{t}$:

$$\sigma_D = \sqrt{\frac{\sigma_1^2}{N_1} + \frac{\sigma_2^2}{N_2}} = \sqrt{\frac{(25)^2}{6} + \frac{(40)^2}{5}} = \sqrt{104 + 320}$$

$$= 20.6$$

$$\bar{t} = \frac{\bar{x}_1 - \bar{x}_2}{\sigma_D} = \frac{15 - 7}{20.6} = 0.39$$

degrees of freedom $= N_1 + N_2 - 2 = 6 + 5 - 2 = 9$

t for 95% confidence $= 2.26$

Note that increasing the dispersion of the sample (increasing the standard deviation) caused a decrease in the value of $\bar{t}$ but did not influence the critical values of t. This means that there is a greater chance that the samples had some elements in common. Hence it becomes more difficult to state that they are not from the same or similar populations. If the sample means were much different, say $\bar{x}_1 = 5$ and $\bar{x}_2 = 150$, then there would be less chance of the values from one group being also included in the other group and a greater possibility of the populations not being similar. This is reflected in the increased value of $\bar{t}$. Holding the other variables constant, we obtain

$$\bar{t} = \frac{\bar{x}_1 - \bar{x}_2}{\sigma_D} = \frac{5 - 150}{20.6} = -7.04$$

-7.04 definitely lies within the reject region.

Example 9.5

A researcher wants to determine if there is a difference between Republicans and Democrats with respect to authoritarianism. She takes a random sample from each group and finds the authoritarianism scores for the Democrats and Republicans, respectively, to be the following:

Democrats	Republicans
$\bar{x}_D = 7.1$	$\bar{x}_R = 9.9$
$\sigma_D = 8$	$\sigma_R = 10$
$N_D = 32$	$N_R = 50$

With 99% confidence, decide if there is a significant difference. To start, hypothesize that there is *not* a difference.

$$H_0: \quad \mu_D = \mu_R$$
$$H_i: \quad \mu_D \neq \mu_R$$

Then calculate σ_D and $\bar{t}$:

$$\sigma_D = \sqrt{\frac{\sigma_1{}^2}{N_1} + \frac{\sigma_2{}^2}{N_2}} = \sqrt{\frac{8^2}{32} + \frac{10^2}{50}} = \sqrt{2 + 2}$$

$$= 2$$

$$\bar{t} = \frac{\bar{x}_1 - \bar{x}_2}{\sigma_D} = \frac{7.1 - 9.9}{2}$$

$$= -1.4$$

For the critical value of t we have

degrees of freedom $= N_1 + N_2 - 2 = 32 + 50 - 2 = 80$

t for 99% confidence $= 2.56$

The decision rule is, if $\bar{t}$ is greater than -2.56 and less than 2.56, accept the hypothesis. If $\bar{t}$ is less than -2.56 or greater than 2.56 then, reject the hypothesis.

The $\bar{t}$ value of -1.4 indicates that the hypothesis should be accepted.

PROBLEMS

Problem 9.1

Samples are taken of a set of workers at two different times to see what effect fatigue has on their output. The sample at both times consisted of 81 people. At time 1 they produced 50 units/min, with a standard deviation of 5 units/min. At time 2 they produced 40 units/min, with a standard deviation of 15 units/min. With 95% confidence, can one say that there is a significant difference in their outputs?

$H_0:$ There is no change (fatigue not a factor), $\mu_1 = \mu_2$.
$H_i:$ There is a change, $\mu_1 = \mu_2$.

Then calculate σ_D and $\bar{t}$:

$$\sigma_D = \sqrt{\frac{\sigma_1{}^2}{N_1} + \frac{\sigma_2{}^2}{N_2}} = \sqrt{\frac{(5)^2}{81} + \frac{(15)^2}{81}} = \sqrt{\frac{25 + 225}{81}}$$

$$= 1.76$$

$$\bar{t} = \frac{\bar{x}_1 - \bar{x}_2}{\sigma_D} = \frac{50 - 40}{1.76}$$

$$= 5.68$$

degrees of freedom $= N_1 + N_2 - 2 = 81 + 81 - 2 = 160$

For 95% confidence and 160 degrees of freedom, from the t table, we have critical values of $t = \pm 1.96$ (Figure 9.3).

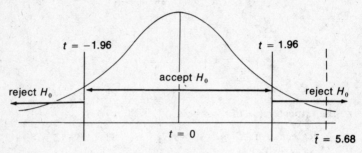

$t = -1.96$ $t = 1.96$

accept H_0

reject H_0 reject H_0

$t = 0$

$\bar{t} = 5.68$

Figure 9.3

Reject H_0 because $\bar{t}$ lies within the reject region.

Problem 9.2

Terri, an office manager, wishes to know whether or not there is a significant difference between the typing speeds of secretaries and typists. Secretaries type 65 words/min, with a variance of 23 words/min, and typists type 75 words/min, with a variance of 31 words/min. Is the difference significant at the 90% confidence level? There are 8 secretaries rated and 6 typists.

H_0: There is no difference, $\mu_S = \mu_T$.
H_i: There is a difference, $\mu_S = \mu_T$.

$$\sigma_D = \sqrt{\frac{\sigma_1^2}{N_1} + \frac{\sigma_2^2}{N_2}} = \sqrt{\frac{23}{8} + \frac{31}{6}}$$

$$= 2.84$$

$$\bar{t} = \frac{\bar{x}_1 - \bar{x}_2}{\sigma_D} = \frac{65 - 75}{2.84}$$

$$= -3.52$$

degrees of freedom $= N_1 + N_2 - 2 = 8 + 6 - 2 = 12$

For 12 degrees of freedom and 90% confidence, the critical t values are ± 1.78 (Figure 9.4).

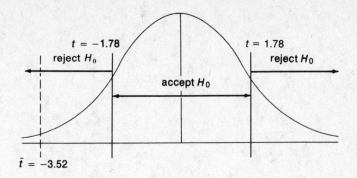

Figure 9.4

Reject H_0 because $\bar{t}$ lies within the reject region.

Problem 9.3

Can you say, with 90% confidence, whether or not the following samples are significantly different?

Sample A	Sample B
10, 12, 14, 15, 14	2, 9, 13, 15, 6

$$H_0: \quad \mu_A = \mu_B$$
$$H_i: \quad \mu_A \neq \mu_B$$

Sample A

$$\bar{x}_A = \frac{1}{N} \sum_{i=1}^{n} x_i\, f(x_i) = \frac{1}{5}\,(10 + 12 + 14 + 15 + 14)$$

$$= 13$$

$$\bar{\sigma}_A{}^2 = \frac{1}{N} \sum_{i=1}^{n} (x_i - \bar{x})^2 f(x_i) = \frac{1}{5} \sum_{i=1}^{n} (x_i - 13)^2 f(x_i)$$

$$= \frac{1}{5}\,[(10 - 13)^2 1 + (12 - 13)^2 1 + (14 - 13)^2 2 + (15 - 13)^2 1]$$

$$= \frac{1}{5}\,(9 + 1 + 2 + 4)$$

$$= 3.2$$

Sample B

$$\bar{x}_B = \frac{1}{N} \sum_{i=1}^{n} x_i f(x_i) = \frac{1}{5} (2 + 9 + 13 + 15 + 6)$$

$$= 9$$

$$\bar{\sigma}_B{}^2 = \frac{1}{N} \sum_{i=1}^{n} (x_i - \bar{x})^2 f(x_i) = \frac{1}{5} \sum_{i=1}^{n} (x_i - 9)^2 f(x_i)$$

$$= \frac{1}{5} [(2 - 9)^2 1 + (9 - 9)^2 1 + (13 - 9)^2 1 + (15 - 9)^2 1$$

$$+ (6 - 9)^2 1]$$

$$= \frac{1}{5} (49 + 0 + 16 + 36 + 9)$$

$$= 22$$

For the difference in means:

$$\sigma_D = \sqrt{\frac{\sigma_A{}^2}{N_A} + \frac{\sigma_B{}^2}{N_B}} = \sqrt{\frac{3.2}{5} + \frac{22}{5}}$$

$$= 2.245$$

$$\bar{t} = \frac{\bar{x}_A - \bar{x}_B}{\sigma_D} = \frac{13 - 9}{2.245}$$

$$= 1.78$$

degrees of freedom $= N_A + N_B - 2 = 5 + 5 - 2 = 8$

For 8 degrees of freedom and 90% confidence, the critical t values are ± 1.86 (Figure 9.5).

So accept the H_0.

Problem 9.4

With 99% confidence, is there a significant difference between the means of the following samples?

Sample A	Sample B
22, 31, 9, 52, 5	5, 10, 15

$$H_0: \quad \mu_A = \mu_B$$
$$H_i: \quad \mu_A = \mu_B$$

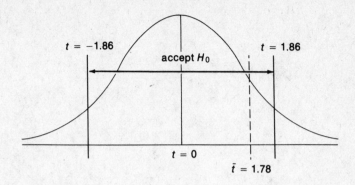

Figure 9.5

Sample A

$$\bar{x}_A = \frac{1}{N} \sum_{i=1}^{n} x_i f(x_i) = \frac{1}{5} (22 + 31 + 9 + 52 + 5)$$

$$= 24$$

$$\sigma_A{}^2 = \frac{1}{N} \sum_{i=1}^{n} (x_i - \bar{x})^2 f(x_i) = \frac{1}{5} \sum_{i=1}^{n} (x_i - 24)^2 f(x_i)$$

$$= \frac{1}{5} [(22 - 24)^2 1 + (31 - 24)^2 1$$

$$+ (9 - 24)^2 1 + (52 - 24)^2 1 + (5 - 24)^2 1]$$

$$= 284$$

Sample B

$$\bar{x}_B = \frac{1}{N} \sum_{i=1}^{n} x_i f(x_i) = \frac{1}{3} (5 + 10 + 15)$$

$$= 10$$

$$\bar{\sigma}_B{}^2 = \frac{1}{N} \sum_{i=1}^{n} (x_i - \bar{x})^2 f(x_i) = \frac{1}{3} \sum_{i=1}^{3} (x_i - 10)^2 f(x_i)$$

$$= \frac{1}{3} \left[(5 - 10)^2 1 + (10 - 10)^2 1 + (15 - 10)^2 1 \right]$$

$$= 16.7$$

For the difference in means:

$$\sigma_D = \sqrt{\frac{\bar{\sigma}_A^2}{N_1} + \frac{\bar{\sigma}_B^2}{N_2}} = \sqrt{\frac{284}{5} + \frac{16.7}{3}}$$

$$= 7.9$$

$$\bar{t} = \frac{\bar{x}_A - \bar{x}_B}{\sigma_D} = \frac{24 - 10}{7.9}$$

$$= 1.77$$

degrees of freedom: $N_1 + N_2 - 2 = 5 + 3 - 2 = 6$

For 6 degrees of freedom and 99% confidence, the critical t values are ± 3.71 (Figure 9.6).

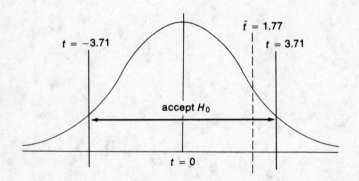

Figure 9.6

Accept H_0 (there is no significant difference between the means).

10

Coefficient of Correlation

The coefficient of correlation is used to determine whether there is some correlation between two factors, such as weight and height or years of experience and salary. Correlation entails the possibility of predicting in some manner what one variable should be if we know the values of the other variable. For instance, if we know the number of years that an engineer has worked, can we predict the salary level? With respect to model building, the coefficient of correlation can be used in the analysis to build the model to make predictions about one variable based on another variable or to compare predictions with occurrences.

If we can predict exactly the change in a second variable when we know the change in a first variable, then the total variation of the second variable has been *accounted for* by the change in the first variable. The square root of the ratio of variation accounted for to the total variation is called the coefficient of correlation (denoted by r).

$$\text{coefficient of correlation} = r = \pm \sqrt{\frac{\text{variation accounted for}}{\text{total variation}}}$$

INTERPRETING THE COEFFICIENT

A coefficient equal to 1 means that there is a perfect correlation; i.e., all the variation has been accounted for and the numerator equals the denominator in the expression for r. An example of this is the side of a square and the area of the square. As a side is increased we can always predict what the area will be, so all the variation in the area is accounted for.

A coefficient of 0 means that there is no correlation between the variables; i.e., knowing the variation in one of the variables does not increase one's ability to predict the possible variations in the other variable, For example, your I.Q. and the population of Cuba have an r of 0.

A positive correlation means that as one variable increases (or decreases) in value, so does the other variable. A negative correlation means that as one variable increases in value, the other variable decreases in value.

The coefficient can be between -1 and $+1$. If your calculations ever yield a coefficient that is less than -1 or greater than $+1$, then you have made a mistake. 0 is no correlation, and moving from 0 towards $+1$ or -1, the correlation gets stronger, until at $+1$ or -1 it is "perfect." A coefficient of 0.5 indicates that half the variation is accounted for by the change in the other variable and is thus predictable. What is a sufficient correlation for the problem solver? As a rule of thumb, anything above 0.9 is almost always considered acceptable, although in some problems the standard might drop much below this.

The coefficient is frequently used to measure how well the line generated by the least squares linear regression technique actually fits the data, because although this technique yields the straight line that best fits the data, there might be no straight line that actually fits the data very well.

Note that r is a dimensionless quantity, so it may be applied to any units of measurement.

The formula for the coefficient is

$$r = \frac{N\Sigma xy - (\Sigma x)(\Sigma y)}{\sqrt{[\Sigma x^2 - (\Sigma x)^2][N\Sigma y^2 - (\Sigma y)^2]}}$$

where

N = number of x's or number of y's (the number of x's must equal the number of y's)

x = an individual number from one of the factors

$(\Sigma x)^2$ = sum of all the individual numbers in factor x squared. Note: add then square.

Σx^2 = sum of all the squared individual numbers in factor x. Note: square then add.

y = an individual number from one of factors (factor y)

$(\Sigma y)^2$ = sum of all the individual numbers in factor y squared. Note: add then square.

Σy^2 = the sum of all the squared individual numbers in factor y

xy = an ordered pair of an element from x multiplied by an element from y

RANK COEFFICIENT

Sometimes it is not possible to assign exact, meaningful numerical quantities to certain variables. The variables that can be assigned values i.e., can be quantified, are usually called *objective*; those that cannot,

COEFFICIENT OF CORRELATION

subjective. Typically, subjective variables can be *ranked*: things or people with the characteristic of the variable can be put in order of increasing incidence of the characteristic. For instance, employees can be ranked on such attributes as compatibility or leadership. It is impossible to assign a score of say 70 or 50 to each employee, but one may put them in order, so that the leader is given a 10, the second-place person a 9, and so forth. If ranking is done for two factors, then the correlation between them can be measured by *Spearman's formula*, which is

$$r_{\text{rank}} = 1 - \frac{6 \, \Sigma D^2}{N \, (N^2 - 1)}$$

where

D = difference between ranks of corresponding values of x and y
N = number of pairs of values of the data

The interpretation of the calculated r value is the same as for the coefficient of correlation. This technique is used in evaluating subjective data that one cannot quantify.

Example 10.1

Find the coefficient of correlation for the following two sets of numbers:

x	y
2	5
7	4
8	1
10	3

The coefficient of correlation is

$$r = \frac{N\Sigma xy - (\Sigma x)(\Sigma y)}{\sqrt{[N\Sigma x^2 - (\Sigma x)^2][N\Sigma y^2 - (\Sigma y)^2]}}$$

We solve in table form for the quantities needed.

	x	y	xy	x^2	y^2
	2	5	10	4	25
	7	4	28	49	16
	8	1	8	64	1
	10	3	30	100	9
Totals	27	13	76	217	51

Substitute into the formula

$$N = 4$$
$$\Sigma xy = 76$$
$$\Sigma x = 27$$
$$\Sigma y = 13$$
$$\Sigma x^2 = 217$$
$$(\Sigma x)^2 = (27)^2 = 729$$
$$\Sigma y^2 = 51$$
$$(\Sigma y^2) = (13)^2 = 169$$

$$r = \frac{4(76) - 27(13)}{\sqrt{[4(217) - 729][4(51) - 169]}}$$

$$= \frac{304 - 351}{\sqrt{(868 - 729)(204 - 169)}}$$

$$= \frac{-47}{\sqrt{(139)(35)}}$$

$$= \frac{-47}{70} = -0.67$$

The negative sign indicates that as x increases, y decreases. If r were -1.00, then there would be a perfect negative correlation, so that if one knew the change in x he or she could exactly predict the change in y. -0.67 indicates a reasonable predictive value for y in terms of x.

Example 10.2

The average salary and years of experience for managers at the Matsuda Company were the following:

Years of experience	1	2	3	4	5
Average salary (in $1000's)	9	11	12	12	13

Is there a correlation between years of experience and salary? The coefficient of correlation is

$$r = \frac{N\Sigma xy - (\Sigma x)(\Sigma y)}{\sqrt{[N\Sigma x^2 - (\Sigma x)^2][N\Sigma y^2 - (\Sigma y)^2]}}$$

We solve in table form for the quantities needed.

	x	y	xy	x^2	y^2
	1	9	9	1	81
	2	11	22	4	121
	3	12	36	9	144
	4	12	48	16	144
	5	13	65	25	169
Totals	15	57	180	55	659

$$N = 5$$
$$\Sigma xy = 180$$
$$\Sigma x = 15$$
$$\Sigma y = 57$$
$$\Sigma x^2 = 55$$
$$(\Sigma x)^2 = (15)^2 = 225$$
$$(\Sigma y)^2 = (57)^2 = 3249$$

$$r = \frac{5(180) - 15(57)}{\sqrt{[5(55) - 225][5(659) - 3249]}}$$

$$= \frac{900 - 855}{\sqrt{[275 - 225][3295 - 3249]}}$$

$$= \frac{45}{\sqrt{(50)(46)}}$$

$$= \frac{45}{48} = 0.938$$

This indicates a very strong correlation between the two variables, years of experience and salary. As one increases, so does the other.

Example 10.3

The sales figures for a local restaurant are shown here with the advertising budget spent for each time period. Are the sales related to the advertising budget?

Advertising budget (in $1000's)	Sales (in $1000's)
10	36
12	14
14	18
10	12
6	4
15	24

Call the advertising figures x and the sales y. The coefficient of correlation is

$$r = \frac{N\Sigma xy - (\Sigma x)(\Sigma y)}{\sqrt{[N\Sigma x^2 - (\Sigma x)^2][N\Sigma y^2 - (\Sigma y)^2]}}$$

We solve in table form for the quantities needed.

x	y	xy	x^2	y^2
10	36	360	100	1296
12	14	168	144	196
14	18	252	196	324
10	12	120	100	144
6	4	24	36	16
15	24	360	225	576
Totals 67	108	1284	801	2552

$$N = 6$$
$$\Sigma xy = 1284$$
$$\Sigma x = 67$$
$$\Sigma y = 108$$
$$\Sigma x^2 = 801$$
$$(\Sigma x)^2 = 4489$$
$$\Sigma y^2 = 2552$$
$$(\Sigma y)^2 = 11,664$$

$$r = \frac{6(1284) - 67(108)}{\sqrt{[6(801) - 4489][6(2552) - 11,664]}}$$

$$= \frac{7704 - 7236}{\sqrt{(4806 - 4489)(15,312 - 11,664)}}$$

$$= \frac{468}{\sqrt{(317)(3648)}} = \frac{468}{1075}$$

$$= 0.44$$

This indicates a positive correlation but not a very strong one, one that would lead us to question the effectiveness of increasing the advertising budget.

Example 10.4

Grant, a consultant on plant locations, wishes to use a sample of 8 locations previously evaluated to know whether or not to promise a favorable location with respect to two subjective factors: quality of the school system and social attitude of the community. If these factors have a high correlation and if this sample is representative, then Grant will know whether or not to promise a high value on both standards.

His sample results show:

Location	A	B	C	D	E	F	G	H
School system	1	2	3	4	5	6	7	8
Social attitude	3	4	1	5	2	7	6	8

To use the rank correlation, the differences must be generated.

Location	A	B	C	D	E	F	G	H
School system	1	2	3	4	5	6	7	8
Social attitude	3	4	1	5	2	7	6	8
Differences (D)	-2	-2	2	-1	3	-1	1	0
D^2	4	4	4	1	9	1	1	0

$$\Sigma D^2 = 24$$

$$r_{\text{rank}} = 1 - \frac{6\Sigma D^2}{N(N^2 - 1)} = 1 - \frac{6(24)}{8[(8)^2 - 1]} = 1 - \frac{144}{8(63)}$$

$$= 1 - 0.29 = 0.71$$

Since this indicates a fairly strong relationship between the two attributes, quality of school system and social attitude, Grant concludes that he should be able to promise a location that ranks high in both attributes.

Example 10.5

The personnel manager wishes to know if there is a correlation between leadership and popularity, with a view to possible modification of the company's orientation program. He has his employees rank 10 managers on both traits. The following table represents the average rank of both traits:

Manager	A	B	C	D	E	F	G	H	I	J
Leadership	1	2	3	4	5	6	7	8	9	10
Popularity	1	3	9	8	5	7	6	2	10	4

How well do they correlate? The ordered pairs can best be evaluated by the rank correlation technique. First the differences between the pairs must be calculated.

Manager	A	B	C	D	E	F	G	H	I	J
Leadership	1	2	3	4	5	6	7	8	9	10
Popularity	1	3	9	8	5	7	6	2	10	4
Differences (D)	0	-1	-6	-4	0	-1	1	6	1	6
D^2	0	1	36	16	0	1	1	36	1	36

$$\Sigma D^2 = 128$$

$$r_{rank} = 1 - \frac{6\Sigma D^2}{N(N^2 - 1)} = 1 - \frac{6(128)}{10[(10)^2 - 1]} = 1 - \frac{768}{10(99)}$$

$$= 1 - 0.78 = 0.22$$

This indicates very little correlation between the two traits, leadership and popularity.

PROBLEMS

Problem 10.1

Connolly Food Company claims that the number of bacteria per gallon of their liquid food decreases as the cooking time is increased. Following are the results of a series of samples:

x = cooking time in minutes
y = number of bacteria per gallon

x	y
2	50
4	45
6	43
8	40
10	37
12	30
14	30
16	30
18	25
20	26

Decide if the two variables are related, and if so, does the number of bacteria per gallon decrease as the cooking time is increased?

The coefficient of correlation is

$$r = \frac{N\Sigma xy - (\Sigma x)(\Sigma y)}{\sqrt{[N\Sigma x^2 - (\Sigma x)^2][N\Sigma y^2 - (\Sigma y)^2]}}$$

Solving for the terms in table form yields

	x	y	xy	x^2	y^2
	2	50	100	4	2500
	4	45	180	16	2025
	6	43	258	36	1849
	8	40	320	64	1600
	10	37	370	100	1369
	12	30	360	144	900
	14	30	420	196	900
	16	30	480	256	900
	18	25	450	324	625
	20	26	520	400	676
Totals	110	356	3458	1540	13,344

$$N = 10$$

$$r = \frac{10(3458) - 110(356)}{\sqrt{[10(1540) - (110)^2][10(13,344) - (356)^2]}}$$

$$= \frac{34,580 - 39,160}{\sqrt{(3300)(6704)}}$$

$$= \frac{-4590}{4703} = -0.97$$

The negative value indicates that as x increases that y decreases. The fact that the coefficient is close to -1.0 indicates that the relationship between x and y is very close. If we know x, we can precisely predict y.

Problem 10.2

A state highway department projects that the average traffic flow for Highway 23 has been and will be the following:

Year	Average daily traffic (in 1000's)
1976	30.0
1980	33.0
1985	37.0
1990	41.0

They wish to determine if there is any correlation between the year and the traffic.

The coefficient of correlation is

$$r = \frac{N\Sigma xy - (\Sigma x)(\Sigma y)}{\sqrt{[N\Sigma x^2 - (\Sigma x)^2][N\Sigma y^2 - (\Sigma y)^2]}}$$

We solve for the values in table form. Note that for ease of calculations, 1976 is treated as the base year 1, so 1980 is year 5, etc.

	x	y	xy	x^2	y^2
	1	30	30	1	900
	5	33	165	25	1089
	10	37	370	100	1369
	15	41	615	225	1681
Totals	31	141	1180	351	5039

$$r = \frac{4(1180) - 31(141)}{\sqrt{[4(351) - (31)][4(5039) - (141)^2]}}$$

$$= \frac{4720 - 4371}{\sqrt{(1404 - 961)(20,156 - 19,881)}}$$

$$= \frac{349}{\sqrt{(443)(275)}}$$

$$= 1$$

The correlation is perfect. If we know a given value for x, we can exactly predict the corresponding value for y. This type of coefficient rarely occurs in actual problems.

Problem 10.3

You work for an advertising agency and handle the Brogyanyi Department Store account. You wish to determine the effect on sales of increasing your advertising budget, so you obtain the following data:

x = thousands of dollars spent on advertising
y = thousands of dollars spent on sales

x	y
10	120
15	140
20	180
25	185

Your question is, what is the correlation, if any, between x and y? You assign the problem to your assistant to do some calculations and he returns the following to you:

x	y	x^2	y^2	$x - y$	$(x - y)^2$
10	120	100	14,400	−110	12,100
15	140	225	19,600	−125	15,625
20	180	400	32,400	−160	25,600
25	185	625	34,225	−160	25,600
Totals 70	625	1350	100,625	−555	78,925

	xy	$(xy)^2$	$y - x$	$(y - x)^2$
	1,200	1,440,000	110	12,100
	2,100	4,410,000	125	15,625
	3,600	12,960,000	160	25,600
	4,625	21,390,625	160	25,600
Totals	11,525	40,200,625	555	78,925

It is now your job to take the calculations done by your assistant and solve the problem of the effect advertising has on sales.

Sometimes it is just as important to know what data you don't need as it is to know what data you do need. In this instance, your assistant did more calculations than were necessary. To decide what is necessary, a good starting point is to list the formula and evaluate the terms within it.

The coefficient of correlation is

$$r = \frac{N\Sigma xy - (\Sigma x)(\Sigma y)}{\sqrt{[N\Sigma x^2 - (\Sigma x)^2][N\Sigma y^2 - (\Sigma y)^2]}}$$

N is 4 because there were 4 sets of ordered pairs (x, y) of data given, and the remainder of terms necessary have already been generated by your assistant.

$$r = \frac{4(11,525) - 70(625)}{\sqrt{[4(1350) - (70)][4(100,625) - (625)^2]}}$$

$$= \frac{46,100 - 43,750}{\sqrt{(5400 - 4900)(402,500 - 390,625)}}$$

$$= \frac{2350}{\sqrt{(500)(11,875)}} = \frac{2350}{2437}$$

$$= 0.96$$

The correlation is very strong and positive. As x is increasing y is also increasing. Thus your conclusion is that extra dollars spent in advertising do assist in increasing sales.

Problem 10.4

Alan evaluates each of the 12 people that he meets by ranking them in alphabetical order, with the number 1 ranking going to a person whose name begins with A. Zeke evaluates each of the same 12 people, but he ranks them in reverse alphabetical order, with the highest rank going to the last person in the alphabet. You are unaware of their respective prejudices and try to evaluate the correlation between their rankings.

Alan's rankings		Zeke's rankings	
Person	Rank	Person	Rank
A	1	A	12
B	2	B	11
C	3	C	10
D	4	D	9
E	5	E	8
F	6	F	7
G	7	G	6
H	8	H	5
I	9	I	4
J	10	J	3
K	11	K	2
L	12	L	1

Evaluate the correlation (rank) between the two rankings.

The best way to approach a ranking problem is to make a single table that reflects both rankings for each person in the same column.

Person	A	B	C	D	E	F	G	H	I	J	K	L
Alan's rankings	1	2	3	4	5	6	7	8	9	10	11	12
Zeke's rankings	12	11	10	9	8	7	6	5	4	3	2	1
Differences D	-11	-9	-7	-5	-3	-1	1	3	5	7	9	11
D^2	121	81	49	25	9	1	1	9	25	49	81	121

$$\Sigma D^2 = 572$$

$$r_{\text{rank}} = 1 - \frac{6\Sigma D^2}{N(N^2 - 1)}$$

$$= 1 - \frac{6(572)}{12(12^2 - 1)} = 1 - \frac{3432}{12(144 - 1)} = 1 - \frac{3432}{1716}$$

$$= -1$$

The value of the coefficient is -1, which means a perfect correlation in the negative sense. This rarely happens in actual practice and it took a specially constructed example here to derive this result. If actual data ever lead to this result, be surprised and hesitant about accepting the result.

Problem 10.5

A manager wishes to know if women's salaries are related to their educational level. She ranks each of 15 women on both scales: salary and education and achieves the following:

Salary rankings		Educational rankings	
Position	Person	Position	Person
1	D	1	B
2	F	2	F
3	E	3	C
4	J	4	H
5	K	5	I
6	L	6	A
7	A	7	D
8	G	8	J
9	I	9	E
10	M	10	M
11	B	11	N
12	H	12	G
13	N	13	L
14	C	14	O
15	O	15	K

Does a high ranking on educational level imply a high salary?

The rank scheme lends itself to using the rank coefficient, for which the formula is

$$r_{rank} = 1 - \frac{6\Sigma D^2}{N(N^2 - 1)}$$

The best way to obtain the D^2 terms is to put the rankings into a table and show both rankings for each person in the same column. This technique is not mandatory, but it certainly simplifies the approach.

Person	A	B	C	D	E	F	G	H	I	J	K	L	M	N	O
Positions	7	11	14	1	3	2	8	12	9	4	5	6	10	13	15
Salaries	6	1	3	7	9	2	12	4	5	8	15	13	10	11	14
Differences (D)	1	10	11	−6	−6	0	−4	8	4	−4	−10	−7	0	2	1
D^2	1	100	121	36	36	0	16	64	16	16	100	49	0	4	1

$$\Sigma D^2 = 560$$

$$r_{rank} = 1 - \frac{6\Sigma D^2}{N(N^2 - 1)} = 1 - \frac{6(560)}{15(15^2 - 1)}$$

$$= 1 - \frac{3360}{15(224)} = 1 - \frac{3360}{3360}$$

$$= 0$$

There is no correlation between the attributes.

11

χ^2 Statistic

The decision maker may wish to determine, with a specified level of confidence, whether or not sample data conform to a projected performance, or in terms of model building, if occurrences confirm predictions. The projected performance may be generated from past performances or theoretical projections. With respect to model building χ^2 (chi square) can be used to compare predictions with occurrences in order to determine the validity of the predictions. The technique used is to propose a null hypothesis (H_0) associated with which we must be able to generate a set of expected values. Based on this null hypothesis or theoretical guess, we calculate a theoretical distribution that will predict the outcome of an event. Then we establish an alternative hypothesis (H_i) that will have to be correct if the null hypothesis is proved not to be correct. We then compare some sample data of the outcome of the event about which the predictions were made (which are called observed data or observed frequencies) with the theoretical predictions (which were based on the null hypothesis) and either accept as correct or reject as false the null hypothesis.

We check the null hypothesis (theoretical prediction) with the actual outcome of the samples (observed frequencies) by means of the χ^2 statistic to see if the actual sample data conform to the predictions (hence verify the predictions). It should be noted that rarely will the sample data exactly conform to the predictions. For a variety of reasons this does not necessarily mean that the predictions are wrong. The key question is, is the difference statistically significant?

The standard form of the χ^2 problem is illustrated in Figure 11.1

The test always has the "accept" region on the left (or less than) the critical value of the χ^2 statistic (obtained from the table) and "reject" on the right (or greater than). The χ^2 for the sample data is calculated and applied "to the test."

To apply the χ^2 statistical technique, the null hypothesis is used to generate a set of values that would occur if the null hypothesis is true.

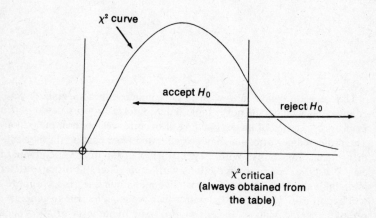

Figure 11.1

These values are referred to as "expected" values. Expected values are put into a rectangular matrix where each cell of the matrix represents a possible outcome. Each expected value is put in parentheses within the cell. The rows and columns of the matrix are established and labeled so that each one represents an outcome or sets of outcomes with common traits. The sample data are then entered into the matrix in the appropriate cells. When constructing a matrix, the entries on a horizontal line are called rows and the vertical entries are called columns. For example, a matrix with 1 row and 2 columns is

	Column 1	Column 2
Row 1		

Figure 11.2

A matrix with 2 rows and 1 column is

Column 1

Row 1

Row 2

Figure 11.3

A matrix with 3 rows and 2 columns is

Column 1   Column 2

Row 1

Row 2

Row 3

Figure 11.4

In Figure 11.2, there are 2 cells. In Figure 11.3, there are 2 cells. In Figure 11.4, there are 6 cells.

An example of how to set up a problem: Suppose you think that a truck driver will drive exactly 6 hours a day for Monday to Friday, so you hypothesize this, with your alternative hypothesis being that this is not true.

Your matrix would be

	M	T	W	Th	F
Driving time	(6)	(6)	(6)	(6)	(6)

This is a matrix with 1 row and 5 columns, and the expected values based on the null hypothesis have been entered into each of the 5 cells. Then your dispatcher tells you that a driver turned in the following for hours driven per day for Monday–Friday of last week: 7, 5, 6, 4, 3. These "observed" values are then entered into the appropriate cells in the matrix in parentheses below the expected values.

	M	T	W	Th	F
Driving time	(6)	(6)	(6)	(6)	(6)
	7	5	6	4	3

Note that when the observed values are entered into the matrix that the number of cells remains constant and the shape of the matrix remains unchanged. Each cell contains an expected value (e) and an observed value (o), each with the same subscript; i.e., for example, the ith cell contains e_i and o_i.

The formula for the χ^2 statistic, where there are cells, is

$$\chi^2 = \sum_{i=1}^{n} \frac{(o_i - e_i)^2}{e_i}$$

To apply this formula, for each cell subtract the expected value from the observed value and square the result. Then divide the squared term by the expected value from the cell. Repeat this for each cell and total the results. This total is the χ^2 statistic for the sample, which is then compared with the value obtained from the table of standard χ^2 values.

In order to solve some problems you will obtain a total for all the items in a row or column and list this amount on the matrix under the heading "total." A total for a row or column is sometimes called a marginal entry or marginal total to the matrix since it is always placed on the edge or margin of the matrix. The values from the cells containing the marginal totals are *not* plugged into the formula to determine the calculated χ^2 value. An example of a marginal total for the row in the above example is

	M	T	W	Th	F	Row total
Driving time	(6) 7	(6) 5	(6) 6	(6) 4	(6) 3	(30) 25

The calculation for χ^2 would be

$$\chi^2 = \sum_{i=1}^{n} \frac{(o_i - e_i)^2}{e_i}$$

$$= \sum_{i=1}^{5} \frac{(o_i - e_i)^2}{e_i} = \frac{(7 - 6)^2}{6} + \frac{(5 - 6)^2}{6}$$

$$+ \frac{(6 - 6)^2}{6} + \frac{(4 - 6)^2}{6} + \frac{(3 - 6)^2}{6}$$

$$= \frac{1}{6} + \frac{1}{6} + \frac{0}{6} + \frac{4}{6} + \frac{9}{6}$$

$$= 2.5$$

If the calculated χ^2 value, for the proper degree of freedom, is greater than the allowed χ^2 values (based on the confidence level and the degrees of freedom) taken from a χ^2 table, then reject the null hypothesis of the appropriate level. If the calculated χ^2 value is less than the one obtained from the χ^2 table, then accept the null hypothesis.

TYPES OF PROBLEMS

All the problems are similar in format and calculations, but differ in the type of null hypothesis used. There are two types of problems:

Type I problems. These, commonly called *contingency table problems*, are applied to see if several characteristics are related. The null hypothesis is always that they are *not* related, the alternative being that they are related.

Type II problems. These problems occur when one wishes to know if a given distribution occurs. The null hypothesis is always that the distribution does occur, the alternative being that it does not occur.

Degrees of freedom and confidence levels for χ^2 statistics are indicated in Appendix III.

TYPE I PROBLEMS (CONTINGENCY TABLES): A situation frequently encountered in many lines of investigation is whether or not two principles of classification are associated. For example, if students in science courses are classified according to grade (below average, average, above average) and to sex, one may wish to know whether there is evidence of a relationship between sex and grade. Or medical researchers may seek to discover if a certain remedy for a sickness actually increases the recovery rate.

The procedure involves classifying sampling units in two ways. For instance, women would be classified according to marital status and degree of friendship participation in order to evaluate the correlation between those two factors.

Example 11.1

Marital status

	Married	Widowed	Totals
High friendship participation	62	83	145
Low friendship participation	28	57	85
Totals	90	140	230

Does the *contingency table* contain evidence that marital status is related to friendship participation? To secure an answer, we argue as follows: Of the 230 women in the sample, 63% have high friendship participation. If there is no relationship between marital status and friendship participation, we would expect to find that 63% of the widows show high friendship participation, and 63% of the married women also show such participation. The hypothesis that there is no relationship between the characteristics leads to the following table for expected values.

	Married	Widowed	Totals
High friendship participation	57	88	145
Low friendship participation	33	52	85
Totals	90	140	230

The value 88 is found by taking 63% of 140; the value 57, by taking 63% of 90; 52 is 37% of 140; 33 is 37% of 90.

Now, do the actual values found in the sample tend to disagree markedly with the expectations? Is 83 really very different from 88, 57 very different from 52, and so forth, or might these differences be attributed to chance? As with other instances of hypothesis testing, we will not be content with vague generalities but will require an answer given with some known risk of making a type I error.

This information can be put in a matrix, with the expected values shown in parentheses, as follows:

	Married	Widowed	Totals
High friendship participation	62 (57)	83 (88)	145
Low friendship participation	28 (33)	57 (52)	85
Totals	90	140	230

We list the actual results alongside the expectations:

Classification combination	Observed value	Expected value	Discrepancy
Married–high	62	57	5
Married–low	28	33	−5
Widowed–high	83	88	−5
Widowed–low	57	52	5

A reasonable person would expect that the greater the discrepancies among the results, the smaller the probability of there being no association. It turns out this is true, but a judgment is not reached on the basis of the absolute size of the discrepancies; instead, we compute χ^2 for the reason that we have tables of a distribution useful in reaching our decision if we do so. The value of χ^2 is found by summing the square of each discrepancy divided by its expectation:

$$\chi^2 = \sum_{i=1}^{n} \frac{o_i - e_i}{e_i}$$

Here o is the observed value, e is the expected value, and the subscript i indicates that we take the values for each term in turn. For this example we have

$$\chi^2 = \frac{5^2}{57} + \frac{(-5)^2}{33} + \frac{(-5)^2}{88} + \frac{5^2}{52} = 1.95$$

We can readily see that if the discrepancy were larger, say, 8 instead of 5, the value of χ^2 would be larger. Therefore, large values of χ^2 result when discrepancies are large (and large discrepancies cast doubt on the idea that the classifications are independent).

One may inquire how large χ^2 must be to furnish evidence of an association. That is where the tables of χ^2 serve a useful purpose. These tables contain values of χ^2 at the 95 percentile, 99 percentile, etc., and may be used to find a "critical" value of χ^2 for a significance test. To find the correct critical value one must set the α-risk and determine the degrees of freedom.

An α-risk is the possibility of making an error and is equal to 1 minus the percent confidence that is listed in the table as a confidence level.

For a contingency table the degrees of freedom are found by multiplying $(c - 1)$ by $(r - 1)$, where c = number of columns in the table and r = number of rows. In the above example $(c - 1)(r - 1) = (2 - 1)(2 - 1) = 1$. If α is set at 5%, the critical χ^2 is 3.84 (see the χ^2 table, Appendix III). Consequently χ^2 of 1.95 is not large enough to furnish evidence of a relationship between marital status and friendship participation. (See Figure 11.5).

Note that if we multiply the *marginal totals* for a given classification combination and divide this product by the total sample size, we have the "expectation" for that classification combination: $140 \times 145 \div 230 = 88$; $90 \times 145 \div 230 = 57$, etc. This is the usual way of computing expectations.

In our 2 by 2 matrix, once an expectation has been computed (such as 88 above), the others can be found by subtraction: $145 - 88 = 57$;

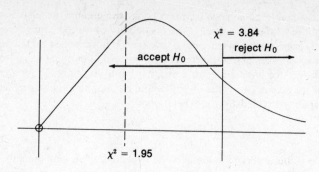

Figure 11.5

$90 - 57 = 33$; $85 - 33 = 52$. In any contingency table after $(c - 1)(r - 1)$ expected values have been computed the remainder may be found by subtraction.

The sum of discrepancies, $\Sigma(o - e)$ is 0. This is not a foolproof check against error, but if your sum is different from 0, you have an error in your work at some stage.

In order to use the χ^2 tables one must have samples with expected values of at least 5.

Example 11.2

A company wished to determine whether absenteeism was related to age of employee. A sample of 200 employees selected at random was classified as follows:

	Age			
Cause of absence	Under 25	25 to 50	Over 50	Totals
Sickness	45	30	45	120
Other	20	35	25	80
Totals	65	65	70	200

Is there evidence, at the 5% significance level, that age and absenteeism are related?

The null hypothesis is age is unrelated to absenteeism. Under this hypothesis, since 60% of the sample was absent because of sickness, we would expect that the 60% ratio to be the same for all age brackets: 60% of 65, 60% of 70, etc.; or we could work with the marginal totals: 65 × 120 ÷ 200, etc.

The table, with expected frequencies shown in parentheses, is as follows:

Cause of absence	Under 25	25 to 50	Over 50	Totals
Sickness	45 (39)	30 (39)	45 (42)	120
Other	20 (26)	35 (26)	25 (28)	80
Totals	65	65	70	200

We list the discrepancies:

	Observed value	Expected value	Discrepancy
Sickness:			
under 25	45	39	6
25 to 50	30	39	−9
over 50	45	42	3
Other:			
under 25	20	26	−6
25 to 50	35	26	9
over 50	25	28	−3

The degrees of freedom is $(3 - 1)(2 - 1) = 2$, and the critical χ^2 from the table is 5.99. We compute χ^2.

$$\chi^2 = \frac{6^2}{39} + \frac{(-9)^2}{39} + \frac{3^2}{42} + \frac{(-6)^2}{26} + \frac{9^2}{26} + \frac{(-3)^2}{28}$$

$$= 8.03$$

The computed χ^2 value is greater than the critical χ^2 value taken from the table, so we conclude that absenteeism and age are related. (See Figure 11.6.)

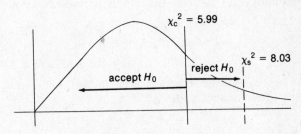

Figure 11.6

Example 11.3

A study was made of the amount of defective work produced by shift in a given plant with the following results:

Shift	Defective pieces	Good pieces	Totals
8:00–4:00	50	920	970
4:00–12:00	60	900	960
12:00–8:00	70	840	910
Totals	180	2660	2840

Can one with 95% confidence conclude the shift has an effect on defective work?

$$H_0: \text{ the shift does not have an effect.}$$
$$H_i: \text{ the shift has an effect.}$$

We compute the expected values. If the shift has no effect, then the proportion of defective items in any shift will be the same as the overall proportion of defective items.

The expected number of defective items for the 8–4 shift is

$$e_{8-4}^{\text{def}} = (\text{proportion defective in total})(\text{number produced 8–4})$$

$$= \frac{180}{2840}(970) = 62$$

The expected number defective for the 4–12 shift is

$$e_{4-12}^{\text{def}} = (\text{proportion defective in total})(\text{number produced 4–12})$$

$$= \frac{180}{2840}(960) = 61$$

To find the 12–8 expected number defective: the total expected defective items must equal the total that actually occurred, namely, 180, so

$$e_{8-4}^{\text{def}} + e_{4-12}^{\text{def}} + e_{12-8}^{\text{def}} = 180$$

$$e_{12-8}^{\text{def}} = 180 - e_{8-4}^{\text{def}} - e_{4-12}^{\text{def}}$$

$$= 180 - 62 - 61 = 57$$

The expected number of good items in the 8–4 shift must equal the total that occurred, namely, 970, so

$$e_{8-4}^{\text{def}} + e_{8-4}^{\text{good}} = 970$$

$$e_{8-4}^{\text{good}} = 970$$

$$- e_{8-4}^{\text{def}} = 908$$

The expected number of good items in the 4–12 shift must equal the total that occurred, namely, 960, so

$$e_{4-12}{}^{\text{def}} + e_{4-12}{}^{\text{good}} = 960$$
$$e_{4-12}{}^{\text{good}} = 960$$
$$- e_{4-12}{}^{\text{def}} = 899$$

The expected number of good items in the 4–12 shift must equal the total that occurred, namely, 910, so

$$e_{12-8}{}^{\text{def}} + e_{12-8}{}^{\text{good}} = 910$$
$$e_{12-8}{}^{\text{good}} = 910$$
$$- e_{12-8}{}^{\text{def}} = 853$$

The completed matrix is (with expected values in parentheses)

Shift	Defective pieces	Good pieces	Totals
8:00–4:00	50 (62)	920 (908)	970
4:00–12:00	60 (61)	900 (899)	960
12:00–8:00	70 (57)	840 (853)	910
Totals	180	2660	2840

We calculate the χ^2 statistic:

$$\chi^2 = \Sigma \frac{(o - e)^2}{e} = \frac{(50 - 62)^2}{62} + \frac{(60 - 61)^2}{61} + \frac{(70 - 57)^2}{57}$$

$$+ \frac{(920 - 908)^2}{908} + \frac{(900 - 899)^2}{899} + \frac{(840 - 853)^2}{853}$$

$$= 2.32 + 0.02 + 2.96 + 0.16 + 0.000 + 0.20$$

$$= 5.66$$

The degrees of freedom is $(r - 1)(c - 1) = (3 - 1)(2 - 1) = 2$, and at 95% confidence, $\chi^2 = 5.99$.

The calculated value is 5.66, which is less than the table value of 5.99, so the null hypothesis, that the shift has no effect on the defective work, cannot be rejected. However, if this were an actual situation, the prudent researcher would take more samples because of the small difference between the calculated value and the value from the table. (See Figure 11.7.)

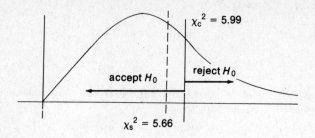

$$\chi_c^2 = 5.99$$

accept H_0

reject H_0

$$\chi_s^2 = 5.66$$

Figure 11.7

Example 11.4

From a random sample, voters (100 with low incomes, 140 with middle incomes, and 60 with high incomes) are asked how they would vote on a certain piece of legislation. The results are shown in the following table.

	Voters' incomes (observed)			
	Low	Middle	High	Totals
For the legislation	58	77	27	162
Against the legislation	42	63	33	138
Totals	100	140	60	300

Are the actual proportions of favorable votes the same for all three groups? In other words, can the sample differences be attributed to chance? Denote the proportions of voters favoring the legislation in the three income groups as p_1, p_2, and p_3.

$$H_0: \quad p_1 = p_2 = p_3$$
$$H_i: \quad p_1, p_2, \text{ and } p_3 \text{ are not all equal}$$

If the null hypothesis is true, we can combine the three samples and estimate the common proportion of voters favoring the legislation as

$$\frac{58 + 77 + 27}{100 + 140 + 60} = \frac{162}{300} = 0.54$$

With this estimate we would *expect* for the legislation 100(0.54) = 54 votes in the first sample, 140(0.54) = 75.6 votes in the second sample, and 60(0.54) = 32.4 votes in the third sample. Subtracting this figure from the respective sample sizes will yield the number of votes *expected* against the legislation.

These results are summarized in the following table, where the *expected frequencies* are shown in the parentheses below those that were actually observed:

	Low	Middle	High	Totals
For the legislation	58 (54)	77 (75.6)	27 (32.4)	162
Against the legislation	42 (46)	63 (64.4)	33 (27.6)	138
Totals	100 (100)	140 (140)	60 (60)	300

To test the null hypothesis, $p_1 = p_2 = p_3$, compare the observed frequencies with the frequencies expected if the null hypothesis were true. For the comparison the χ^2 statistic is used, which is

$$\chi^2 = \sum_{j=1}^{n} \frac{(o_j - e_j)^2}{e_j}$$

where o_j = observed frequency in cell j
e_j = expected frequency in cell j

The equation gives a term for each cell and the sum for the total number of cells.

$$\chi^2 = \frac{(58 - 54)^2}{54} + \frac{(77 - 65.6)^2}{75.6} + \frac{(27 - 32.4)^2}{32.4}$$

$$+ \frac{(42 - 46)^2}{46} + \frac{(63 - 64.4)^2}{64.4} + \frac{(33 - 27.6)^2}{27.6} = 2.656$$

If there is close agreement between the o's and e's, then the difference terms of $(o - e)$, and hence χ^2, will be relatively small. Consequently, we reject the null hypothesis if χ^2 is large and we accept the null hypothesis or reserve judgment if χ^2 is small. The exact value at which we decide whether to accept or reject is based on the level of significance and the degrees of freedom.

The degrees of freedom is $(r - 1)(c - 1) = (2 - 1)(3 - 1) = 2$, and for 95% confidence, we have $\chi^2 = 5.99$

Since 2.656, the calculated value for χ^2, is less than 5.99 obtained from the table, the null hypothesis cannot be rejected. We conclude that income is not a factor in voters' opinion concerning the legislation. (See Figure 11.8.)

TYPE II PROBLEMS: These problems occur when one wishes to know whether a particular distribution occurs. The null hypothesis is that it does occur and will be used in the calculation of the set of expected values.

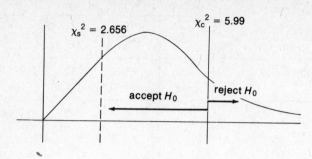

Figure 11.8

Example 11.5

A personnel manager argues that the number of hours of overtime worked per week is not related to the season of the year; i.e., there is no significant difference between the number of overtime hours in the various seasons. Dr. A is hired from Applied Psychological Services to evaluate this claim. How should she proceed? The observed average number of hours per week worked for each season are

Winter	Spring	Summer	Fall
11	15	10	12

First, Dr. A must propose a hypothesis that will enable her to generate a set of expected values for each season. If she proposes that there is a difference among the seasons, she must quantify the difference, that is, she would have to specify exactly what she thinks the differences are. She would rather not do this, so she proposes that there is no difference or that the uniform distribution is applicable.

H_0: The uniform distribution applies.
H_i: The uniform distribution does not apply.

Use H_0 to generate a set of expected values. If the overtime is approximately equal for all seasons, then each one should approximately equal the average value for all of them.

$$\text{mean} = \frac{\Sigma x}{N} = \frac{11 + 15 + 10 + 12}{4} = \frac{48}{4} = 12$$

The observed numbers of hours and, in parentheses, the expected values are indicated:

Winter	Spring	Summer	Fall
11	15	10	12
(12)	(12)	(12)	(12)

Now we calculate the χ^2 statistic:

$$\chi^2 = \sum_{j=1}^{n} \frac{(o_j - e_j)^2}{e_j}$$

$$= \frac{(11 - 12)^2 + (15 - 12)^2 + (10 - 12)^2 + (12 - 12)^2}{12}$$

$$= \frac{14}{12}$$

$$= 1.1667$$

With 95% confidence (the 0.05 level of significance) and with degrees of freedom equal to $(N - 1) = 4 - 1 = 3$, the critical value from the table is 7.82. The calculated χ^2 statistic of 1.1667 is less than the table value of 7.82, so we accept the hypothesis—that the average number of hours worked per week is the same for every season. (See Figure 11.9.)

Example 11.6

The inspector thinks that the width of a part from the lathe operation increases linearly with the increase in the number of minutes spent working it. He proposes the relationship is $y = 2x + 1$

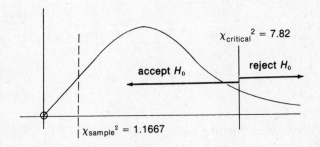

Figure 11.9

where y = width of the part
       x = minutes spent working on it

To substantiate his claim he takes samples which show the following:

x (min)	1	2	3	4	5
y (width)	2	4	10	7	10

He must propose a null hypothesis that will enable him to generate a set of expected values that will be compared with those observed.

H_0: $y = 2x + 1$ represents the relationship.
H_i: $y = 2x + 1$ does not represent the relationship.

For the expected values we have

$$
\begin{aligned}
x &= 1 & y &= 2x + 1 = 2 + 1 = 3 \\
x &= 2 & y &= 4 + 1 = 5 \\
x &= 3 & y &= 6 + 1 = 7 \\
x &= 4 & y &= 8 + 1 = 9 \\
x &= 5 & y &= 10 + 1 = 11
\end{aligned}
$$

In table form (again expected values in parentheses):

x	1	2	3	4	5
y	2	4	10	7	10
	(3)	(5)	(7)	(9)	(11)

$$
\chi^2 = \sum_{j=1}^{n} \frac{(o_j - e_j)^2}{e_j} = \frac{(2 - 3)^2}{3} \quad \frac{(4 - 5)^2}{5}
$$

$$
\times \frac{(10 - 7)^2}{7} \quad \frac{(7 - 9)^2}{9} \quad \frac{(10 - 11)^2}{11}
$$

$$
= \frac{1}{3} + \frac{1}{5} + \frac{9}{7} + \frac{4}{9} + \frac{1}{11}
$$

$$
= 2.35
$$

The critical value from the table, for $N - 1 = 4 - 1 = 3$ degrees of freedom and at the 99% confidence level, is 11.35. The calculated value of 2.35 is less than the table value of 11.35, so we accept the hypothesis. (See Figure 11.10.)

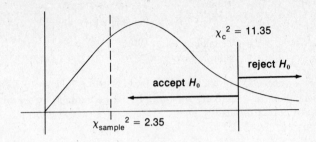

Figure 11.10

PROBLEMS

Problem 11.1

An HEW official decides to test whether income is related to sex for workers of comparable education and experience. She presents you with the following observed data and asks for your conclusion (at the 95% confidence level).

Income (in $1000's)	0–8	8–12	12–16	16–20	More than 20
Male	23	16	25	30	11
Female	36	30	25	20	4

H_0: Sex is not related to income.
H_i: Sex is related to income.

The H_0 is used to generate a set of expected values. There were 105 males tested and 115 females tested, so 220 persons were observed. The column totals are the following for the various income brackets.

Income (in $1000's)	0–8	8–12	12–16	16–20	More than 20
Male	23	16	25	30	11
Female	36	30	25	20	4
Totals	59	46	50	50	15

If sex is not related to income, then we would expect the same proportion of male to female or total persons to be represented in each income bracket as is represented by the sample in general.

$$\frac{\text{males}}{\text{sample size}} = \frac{105}{220} = 0.48$$

So we expect 48% of each income bracket to be males. To translate this percentage to a number, we multiply the percentage by the number of people in each bracket.

	0–8
Males	23
Females	36
Total	59

males in (0–8) = (percentage of males in total sample) (persons in 0–8)

$$= 0.48(59) = 28$$

$$\text{males in } (8–12) = 0.48(46) = 22$$

$$\text{males in } (12–16) = 0.48(50) = 24$$

$$\text{males in } (16–20) = 0.48(50) = 24$$

$$\text{males (over 20)} = 0.48(15) = 7$$

To obtain the expected number of females in each bracket, we can either go through a similar set of calculations by using 0.52 for the percentage of females or use an alternative and shorter way such as the following:

There were 59 people making 0–8, and we expected 28 of them to be males, so we expect 59 − 28 = 31 to be females. Similarly,

$$\text{females in } (8–12) = 46 − 22 = 24$$

$$\text{females in } (12–16) = 50 − 24 = 26$$

$$\text{females in } (16–20) = 50 − 24 = 26$$

$$\text{females in (over 20)} = 15 − 7 = 8$$

In table form we now have the following (the expected values are in parentheses):

Income (in $1000's)	0–8	8–12	12–16	16–20	More than 20
Males	23 (28)	16 (22)	25 (24)	30 (24)	11 (7)
Females	36 (31)	30 (24)	25 (26)	20 (26)	4 (8)
Totals	59	46	50	50	15

Now to make the comparison the χ^2 statistic must be calculated.

$$\chi^2 = \sum_{j=1}^{n} \frac{(o_j - e_j)^2}{e_j}$$

$$= \frac{(23 - 28)^2}{28} + \frac{(16 - 22)^2}{22} + \frac{(25 - 24)^2}{24} + \frac{(30 - 24)^2}{24}$$

$$+ \frac{(11 - 7)^2}{7} + \frac{(36 - 31)^2}{31} + \frac{(30 - 24)^2}{24} + \frac{(25 - 26)^2}{26}$$

$$+ \frac{(20 - 26)^2}{26} + \frac{(4 - 8)^2}{8}$$

$$= 0.9 + 1.7 + 0 + 1.5 + 2.3 + 0.8 + 1.5 + 0 + 1.4 + 2$$

$$= 12.1$$

This calculated value must be compared with a value from the table. The degrees of freedom equals $(r - 1)(c - 1) = (2 - 1)(5 - 1) = (1)(4) = 4$. χ^2 from the table for 95% confidence and 4 degrees of freedom is 9.49. The sample χ^2 (the calculated one) is 12.1. If 12.1 is within the accept region (less than 9.49), then the H_0 must be accepted. If the calculated χ^2 is in the reject region (greater than 9.49), then the hypothesis must be rejected and the alternative hypothesis accepted.

The sample χ^2 is clearly greater than 9.49, so the hypothesis that sex is not related to income must be rejected and one must conclude, based on this sample, that sex is related to income.

Problem 11.2

A transportation manager thinks that the time spent in flying from Philadelphia to New York is always approximately the same. Based on his observations from 5 consecutive days, do you agree (with 95% confidence)?

Day	1	2	3	4	5
Air time (min)	32	40	18	25	35

H_0: The time is always approximately the same.
H_i: The time varies.

This way H_0 can be used to generate a set of expected values, namely,

$$\text{air time/day} = \frac{\text{total air time}}{\text{number of days}}$$

$$= \frac{32 + 40 + 18 + 25 + 35}{5}$$

$$= \frac{150}{5}$$

$$= 30$$

If there is no difference in the time per day, then all the days would require the average time, or 30 min, so one would *expect* this average time per day if the H_0 is correct. The expected values are in parentheses.

Day	1	2	3	4	5
Air time	32 (30)	40 (30)	18 (30)	25 (30)	35 (30)

$$\chi^2 = \sum_{j=1}^{n} \frac{(o_j - e_j)^2}{e_j}$$

$$= \frac{(32 - 30)^2}{30} + \frac{(40 - 30)^2}{30} + \frac{(18 - 30)^2}{30} + \frac{(25 - 30)^2}{30}$$

$$+ \frac{(35 - 30)^2}{30}$$

$$= \frac{4}{30} + \frac{100}{30} + \frac{144}{30} + \frac{25}{30} + \frac{25}{30}$$

$$= \frac{298}{30}$$

$$= 9.93$$

This calculated value must be compared with a value from the table. The degrees of freedom equals $N - 1 = 5 - 1 = 4$. χ^2 from the table for 4 degrees of freedom and 95% confidence is 9.49. Consequently, if the calculated χ^2 is greater than 9.49, we reject H_0; if it is less than 9.49, we accept H_0. The calculated χ^2 is 9.93, which is greater than 9.49, so we reject H_0 (that the time is always approximately the same) and accept H_i (that the time varies).

Problem 11.3

A certain cold remedy, thought to be effective, was tested. In an experiment with 174 persons, half were given the remedy and half were given sugar pills. The subjects' reactions are given in the table:

Effect	Helped	Harmed	No change	Totals
Remedy	56	10	21	87
Sugar pill	46	14	27	87
Totals	102	24	48	174

Is there any evidence at the 5% significance level (95% confidence level) that the remedy is any different from the sugar pill?

H_0: There is no difference between remedy and sugar pill.
H_i: There is a difference.

For the "helped" expected value (using the short-cut rule), we have

$$\frac{\text{row total (column total)}}{\text{grand total}} = \frac{87(102)}{174} = 51$$

For the "harmed" expected value we have

$$\frac{\text{row total (column total)}}{\text{grand total}} = \frac{87(102)}{174} = 12$$

Noting that the expected totals must equal the observed totals, we obtain the other values by subtraction.

Effect	Helped	Harmed	No change	Totals
Remedy	51	12	$87 - (51 + 12) = 24$	87
Sugar pill	$102 - 51 = 51$	$24 - 12 = 12$	$48 - 24 = 24$	87
Totals	102	24	48	174

The completed matrix is (with expected values in parentheses):

Effect	Helped	Harmed	No change	Totals
Remedy	56 (51)	10 (12)	21 (24)	87
Sugar pill	46 (51)	14 (12)	27 (24)	87
Totals	102	24	48	174

We calculate the χ^2 statistic.

$$\chi^2 = \sum_{j=1}^{n} \frac{(o_j - e_j)^2}{e_j} = \frac{(56 - 51)^2}{51} + \frac{(10 - 12)^2}{12} + \frac{(21 - 24)^2}{24}$$

$$+ \frac{(46 - 51)^2}{51} + \frac{(14 - 12)^2}{12} + \frac{(27 - 24)^2}{24}$$

$$= \frac{25}{51} + \frac{4}{12} + \frac{9}{24} + \frac{25}{51} + \frac{4}{12} + \frac{9}{24}$$

$$= 2.40$$

For the critical value from the table, the degrees of freedom equals $(r - 1)(c - 1) = (2 - 1)(3 - 1) = 2$; at 95% confidence, $\chi^2 = 5.99$. The calculated value of 2.40 is less than the table value of 5.99, so the null hypothesis (that there is no difference) should be accepted.

12

Least Squares: Linear Regression

Sometimes we wish to start with data and then derive an equation of a curve that best represents or best "fits" the data. Depending on the shape that the graphing of the data takes, there are many possible curves that might yield the best fit—parabolas, hyperbolas, circles, etc. How well a curve actually fits the data can be analyzed using the coefficient of correlation (Chapter 10). Within this presentation only the most common one, the straight line, will be considered. This assumption of a straight line is important to note.

Once the equation of the line is obtained, there are many practical uses for it. One of the most valuable is to predict values for the dependent variables that correspond to various known or specified values of the independent variable.

Every straight line has an equation of the form $y = a + bx$, so the problem is one of determining the appropriate a and b for a given set of data. Given any two points, one can connect them with a straight line. Given many points, there are many possible straight lines that can connect two or more of the data points. The object is to find the equation of the straight line that will best represent the given data by being closest to all the points, that is, the one where the squared distances (so as to eliminate negative numbers) from the points to the line are at a minimum—hence the name "least squares." (See Figure 12.1.)

For each value of x, the error between the value measured from the sample (y) and the value calculated ($\hat{y}$) from the derived equation $\hat{y} = a + b\hat{x}$) is given by $\bar{y} - (a + b\hat{x})$. The sum of the squares of these errors is what must be minimized. To minimize it, take the derivative with respect to each of the constants (a and b); there are many possible values for a and b and we want only those that minimize the squared distances. To find these, set the derivatives equal to 0 and solve for the values of a and b.

Once the a and b values are obtained, substitute them into the general form of $y = a + bx$, and then the equation of the line is complete. Note

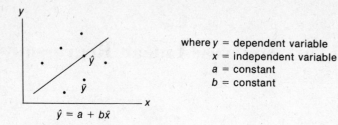

where y = dependent variable
$\quad\quad\;\;\; x$ = independent variable
$\quad\quad\;\;\; a$ = constant
$\quad\quad\;\;\; b$ = constant

Figure 12.1

that although this is the equation of the straight line that best fits the data, perhaps no straight line fits the data very well. (To determine how well the line fits the data see Chapter 10 on correlation.)

An equation in the form $y = a + bx$, with unique values for a and b, is the result. Where $x = 0$ the equation becomes $y = a$, so a is where the line crosses or intercepts the y axis and is referred to as the y intercept. An a value of 0 or negative is a valid result.

In the equation b is the slope of the line; i.e., for a unit change in the horizontal (x) axis, b states the amount of change in the vertical (y) axis. A 0 value for b indicates that y does not depend on x; thus the line is a horizontal line with y constant for every value of x. A negative value for b indicates that as x increases, y decreases. A value of 1 for b indicates that for every increase or decrease in x, y increases or decreases the same amount (see Figure 12.2).

The following formulas give the values of a and b in the equation $y = a + bx$:

$$a = \frac{(\Sigma y)(\Sigma x^2) - (\Sigma x)(\Sigma xy)}{n\Sigma x^2 - (\Sigma x)^2}$$

$$b = \frac{n\Sigma xy - (\Sigma y)(\Sigma x)}{n\Sigma x^2 - (\Sigma x)^2}$$

Note that the denominators are the same for both a and b. The symbols represent the following:

$\quad\quad\;\; n$ = number of x's or number of y's (the number of x's must equal the number of y's)
$\quad\quad\;\; x$ = an individual number from one of the factors
$\;\;(\Sigma x)^2$ = the sum of all the individual numbers in factor x squared; note: add then square
$\quad\;\; \Sigma x^2$ = the sum of all the squared individual numbers in factor x; note: square then add
$\quad\quad\;\; y$ = an individual number from factor y

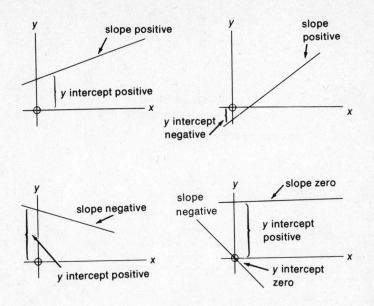

Figure 12.2

xy = an ordered pair of an element from x multiplied by an element from y

Least squares can be used wherever we have two variables. Of course a least squares straight line is the estimate of the curve that best fits the data only when the variables have a simple relation to each other. (Examples include height and weight, years experience and salary.) The coefficient of correlation, as discussed in Chapter 10, is used to determine the reliability of the least squares estimate.

The values for a and b, once calculated by the above formulas, are inserted into $y = a + bx$, and a linear equation is obtained. This equation can be used to predict, for a given x value, what the corresponding y value should be. When graphing or listing the data, the variable measured on the horizontal axis (usually x) is listed first, followed by the variable measured on the vertical axis (usually y), and for each point this can be listed as an ordered pair of (x, y).

Example 12.1

Obtain the least squares estimate of the straight line for the following points:

x	y
1	1
3	2
4	4
6	4
8	5

x = income (in $1000's)
y = years of experience

See also the points graphed in Figure 12.3.

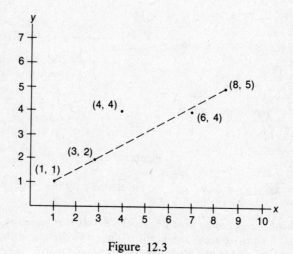

Figure 12.3

The best approach is to use a chart and solve the formulas for a and b.

x	y	xy	x^2
1	1	1	1
3	2	6	9
4	4	16	16
6	4	24	36
8	5	40	64
22	16	87	126

$$y = a + bx$$

$$a = \frac{(\Sigma y)(\Sigma x^2) - (\Sigma x)(\Sigma xy)}{n\Sigma x^2 - n(\Sigma x)^2}$$

$$= \frac{(16)(126) - (22)(87)}{5(126) - (22)^2} = \frac{2016 - 1914}{630 - 484} = \frac{102}{146}$$

$$= 0.69$$

$$b = \frac{n\Sigma xy - (\Sigma x)(\Sigma y)}{n\Sigma x^2 - (\Sigma x)^2} = \frac{5(87) - (22)(16)}{146}$$

$$= \frac{435 - 352}{146} = \frac{83}{146}$$

$$= 0.57$$

$$y = 0.69 + 0.57x$$

To predict what the income should be for a person with 10 years of experience, substitute 10 for x in the equation:

$$y_{(10)} = 0.69 + 0.57(10) = 0.69 + 5.7 = 6.39$$

At $x = 0$, $y = 0.69$, so the line has a y intercept at 0.69. The slope is the b value, or 0.57, so for every increase of 1 unit in x, y increases by 0.57.

Example 12.2

Derive the least squares estimate for the line that best fits the following:

Dollars spent on advertising	x	3	4	5
Income in dollars	y	10	8	11

You want to use x to predict y. To use $y = a + bx$, find a and b.

x	y	xy	x^2
3	10	30	9
4	8	32	16
5	11	55	25
12	29	117	50

$$b = \frac{n\Sigma xy - (\Sigma y)(\Sigma x)}{n\Sigma x^2 (\Sigma x)^2} \quad \frac{3(117) - 29(12)}{3(50) - (12)^2}$$

$$= \frac{351 - 348}{150 - 144} = \frac{3}{6} = 0.5$$

$$a = \frac{(\Sigma y)(\Sigma x^2) - (\Sigma x)(\Sigma xy)}{n\Sigma x^2 - (\Sigma x)^2} = \frac{29(50) - 12(117)}{3(50) - (12)^2}$$

$$= \frac{1450 - 1404}{150 - 144} \quad \frac{46}{6} = 7.67$$

$$y = a + bx = 0.5 + 7.67x$$

Example 12.3

Take the same data in Example 12.2, add the additional items $x = 5$, $y = 9$ and $x = 4$, $y = 12$, derive the equation for the straight line, and use it to predict y for $x = 5$, 100, and 500.

To these

x	y
3	10
4	8
5	11

add these

x	y
5	9
4	12

Put them all in chart form:

x	y	xy	x^2
3	10	30	9
4	8	32	16
5	11	55	25
5	9	45	25
4	12	48	16
21	50	210	91

$$b = \frac{n\Sigma xy - (\Sigma y)(\Sigma x)}{n\Sigma x^2 - (\Sigma x)^2} = \frac{5(210) - 50(21)}{5(91) - (21)^2} = \frac{1050 - 1050}{455 - 441}$$

$$= \frac{0}{14} = 0$$

$$a = \frac{(\Sigma y)(\Sigma x^2) - (\Sigma x)(\Sigma xy)}{n\Sigma x^2 - (\Sigma x)^2} = \frac{50(91) - 21(210)}{14}$$

$$= \frac{4550 - 4410}{14} = \frac{140}{14}$$

$$= 10$$

$$y = a + bx = 10 + 0(x)$$

Use the equation to predict:

$x = 5$	$x = 100$	$x = 500$
$y = 10 + 0(x)$	$y = 10 + 0(x)$	$y = 10 + 0(x)$
$= 10 + 0(5)$	$= 10 + 0(100)$	$= 10 + 0(500)$
$= 10$	$= 10$	$= 10$

Note that the value of y is independent of the value of x; i.e., the prediction is always for $y = 10$ regardless of the x value. This example was put in to demonstrate that in these problems, as with various others, if one of the variables happens to be calculated to be 0, the sample should be checked. There is a chance, as in this case, that for particular numbers a variable is 0. If so, some ridiculous results might occur. This is not to say that every time variables turn out to be 0 a mistake has been made, but it does indicate that we should double check our technique and arithmetic.

Example 12.4

Give the least squares equation for these data and predict y for $x = 19$.

x	y
10	27
15	24
13	26
19	19
21	16

Putting them in chart form, we obtain

x	y	xy	x^2
10	27	270	100
15	24	360	225
13	26	338	169
19	19	361	361
21	16	336	441
78	112	1665	1296

$$b = \frac{n\Sigma xy - (\Sigma y)(\Sigma x)}{n\Sigma x^2 - (\Sigma x)^2} = \frac{5(1665) - 112(78)}{5(1296) - (78)^2}$$

$$= \frac{-411}{396} = -1.04$$

$$a = \frac{(\Sigma y)(\Sigma x^2) - (\Sigma x)(\Sigma xy)}{n\Sigma x^2 - (\Sigma x)^2} = \frac{(112)(1296) - 78(1665)}{396}$$

$$= \frac{15282}{396} = 38.4$$

$$y = 38.4 - 1.04x$$

The negative value for b indicates that as x increases y decreases. Predicting y for $x = 19$, we get

$$y = 38.4 - 1.04(19) = 18.84$$

Note how this compares with the $y = 19$ for $x = 19$ actually measured. Because the line generated need not run exactly through any of the sample points, this is considered a very close prediction for this point (see Figure 12.4).

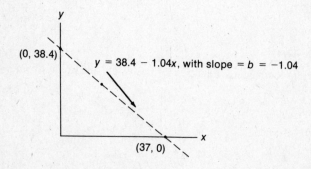

Figure 12.4

The negative slope means that the line is sloped as is shown in the graph.

Example 12.5

Derive the least squares estimate at the straight line that best fits these data:

Height, x	Weight, y
70	155
63	150
72	180
60	135
66	156
70	168
74	178
65	160
62	132
67	145
65	139
68	152

Putting them in chart form and making some calculations, we obtain

Height (in.), x	Weight (lb), y	xy	x^2
70	155	10,850	4,900
63	150	9,450	3,969
72	180	12,960	5,184
60	135	8,100	3,600
66	156	10,296	4,356
70	168	11,760	4,900
74	178	13,172	5,476
65	160	10,400	4,225
62	132	8,184	3,844
67	145	9,715	4,489
65	139	9,035	4,225
68	152	10,336	4,624
802	1850	124,258	53,792

$$ a = \frac{(\Sigma y)(\Sigma x^2) - (\Sigma x)(\Sigma xy)}{n\Sigma x^2 - (\Sigma x)^2} \cdot = \frac{1850(53,792) - 802(124,258)}{12(53,792) - (802)^2} $$

$$= \frac{99,515,200 - 99,654,916}{645,504 - 643,204} = \frac{-139,716}{2,300}$$

$$= -60.7$$

$$b = \frac{n\Sigma xy - (\Sigma y)(\Sigma x)}{n\Sigma x^2 - (\Sigma x)^2} = \frac{12(124,258) - 1850(802)}{12(53,792) - (802)^2}$$

$$= \frac{7396}{2300} = 3.2$$

$$y = a + bx = -60.7 + 3.2x$$

The negative value of a means that at $x = 0$ (the place where the line crosses the y axis, or the y intercept) y is negative. The slope of the line (value of b) is positive (see Figure 12.5).

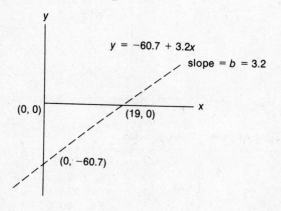

Figure 12.5

PROBLEMS

Problem 12.1

A state highway department projects that the average traffic flow for Highway 23 will be the following:

Year	Average daily traffic (in 1000's of vehicles)
1986	29.7
2000	41.0

You wish to project the traffic for 1990 and 1995. Derive the least squares estimate for the equation of the straight line that best represents the years 1986 and 2000 and then project the traffic for 1990 and 1995.

The general form of the equation of a straight line is always

$$y = a + bx$$

To solve for a and b, the following formulas are used:

$$a = \frac{(\Sigma y)(\Sigma x^2) - (\Sigma x)(\Sigma y)}{n\Sigma x^2 - (\Sigma x)^2}$$

$$b = \frac{n\Sigma xy - (\Sigma y)(\Sigma x)}{n\Sigma x^2 - (\Sigma x)^2}$$

We solve for the variables in table form:

x	y	xy	x^2
1	29.7	29.7	1
15	41	615	225
16	70.7	644.7	226

Note that because a yearly tabulation is desired, if $x = 1$ is for 1986, then $x = 15$ is for 2000.

Plugging into the equations for a and b, we obtain

$$a = \frac{70.7(226) - 16(644.7)}{2(226) - (16)} = \frac{15,978.2 - 10,315.2}{196}$$

$$= 28.9$$

$$b = \frac{2(644.7) - 70.7(16)}{2(226) - (16)} = \frac{1,289.4 - 1,131.2}{196}$$

$$= 0.81$$

The general form of $y = a + bx$ yields

$$y = 28.9 + 0.81x$$

If $x = 1$ represents 1986 and $x = 15$ represents 2000, then for 1990 and 1995 we would have the following:
For 1990, $x = 5$, and the derived formula yields

$$y = 28.9 + 0.81(x) = 28.9 + 0.81(5)$$
$$= 32.95 \text{ projected average daily traffic for 1990}$$

For 1995, $x = 10$

$$y = 28.9 + 0.81x = 28.9 + 0.81(10) = 28.9 + 8.1$$
$$= 37.0 \text{ projected average daily traffic for 1995}$$

Problem 12.2

The number of bacteria per gallon of a liquid food is felt to decrease as the cooking time is increased. The following are the results of a series of samples:

x = cooking time in minutes
y = number of bacteria per gallon

x	y
2	50
4	45
6	43
8	40
10	37
12	30
14	30
16	30
18	25
20	26

Derive the least squares estimate for the straight line that best fits the data and use this estimate to project the number of bacteria that would be expected if the cooking time were 0, 5, 25, and 400 min.

General form of straight line is

$$y = a + bx$$

We need a and b:

$$a = \frac{(\Sigma y)(\Sigma x^2) - (\Sigma x)(\Sigma y)}{n\Sigma x^2 - (\Sigma x)^2}$$

$$b = \frac{n\Sigma xy - (\Sigma y)(\Sigma x)}{n\Sigma x^2 - (\Sigma x)^2}$$

The best way to obtain the data needed for these formulas is to draw a table:

x	y	x^2	xy
2	50	4	100
4	45	16	180
6	43	36	258
8	40	64	320
10	37	100	370
12	30	144	360
14	30	196	420
16	30	256	480
18	25	324	450
20	26	400	520
110	356	1540	3458

$$a = \frac{356(1540) - 110(3458)}{10(1540) - (110)}$$

$$= \frac{548,240 - 380,380}{15,400 - 12,100} = \frac{167,860}{3300}$$

$$= 50.9$$

$$b = \frac{10(3458) - 356(110)}{15,400 - 12,100}$$

$$= \frac{34,580 - 39,160}{3300} = \frac{-4580}{3300}$$

$$= -1.4$$

$$y = a + bx = 50.9 + (-1.4)x = 50.9 - 1.4x$$

The negative sign for b indicates that as x increases, y will decrease; that is, the slope of the line is negative (Figure 12.6).

for $x = 0$
$$y = 50.9 - 1.4(x) = 50.9 - 1.4(0) = 50.9$$

This is where the equation crosses the y axis.

for $x = 5$
$$y = 50.9 - 1.4x = 50.9 - 1.4(5) = 50.9 - 7$$
$$= 43.9$$

for $x = 25$
$$y = 50.9 - 1.4x = 50.9 - 1.4(25) = 50.9 - 35$$
$$= 15.9$$

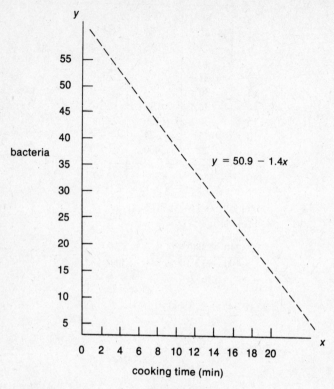

Figure 12.6

for $x = 400$

$$y = 50.9 - 1.4x = 50.9 - 1.4(400) = 50.9 - 560$$
$$= -509.1$$

This result obviously isn't possible (there can never be fewer than 0 bacteria). It was put in to demonstrate that if predictions for extreme values are asked, the problem solver should use extra caution before listing the results. The question is, does this result make sense? Usually there aren't any such problems in a statistics course, but there may be such situations in actual applications on a job.

Problem 12.3

You work for an advertising agency and handle the Wetstein Department Store. You wish to demonstrate the effect on sales of increasing your advertising budget, so you obtain the following data:

x = thousands of dollars spent in advertising
y = thousands of dollars of sales

x	y
10	120
15	140
20	180
25	185

You assign your assistant the task of doing some of the necessary calculations, and he returns the following to you:

x	y	x^2	y^2	$x - y$	$(x - y)^2$
10	120	100	14,400	−110	12,100
15	140	225	19,600	−125	15,625
20	180	400	32,400	−160	25,600
25	185	625	34,225	−160	25,600
70	625	1350	100,625	−555	78,925

xy	$(xy)^2$	$y - x$	$(y - x)^2$
1200	1,440,000	110	12,100
2100	4,410,000	125	15,625
3600	12,960,000	160	25,600
4625	21,390,625	160	25,600
11,525	40,200,625	555	78,925

Derive the least squares estimate for sales in terms of advertising, and use this estimate to project the amount of sales that $30,000 and $35,000 spent on advertising would generate.

Your assistant, attempting to please, has given you too many data in the form of calculations, so it is your task to sort out the unnecessary ones. This is a practical problem. In fact, sometimes the most difficult part of a problem is deciding on the necessary elements. A look at the formulas will prove helpful in deciding what is necessary.

The general form for a straight line is

$$y = a + bx$$

where

$$a = \frac{\Sigma y (\Sigma x^2) - (\Sigma x)(\Sigma xy)}{n\Sigma x^2 - (\Sigma x)^2}$$

$$b = \frac{n\Sigma xy - \Sigma y \Sigma x}{n\Sigma x^2 - (\Sigma x)^2}$$

$$a = \frac{625(1350) - 70(11,525)}{4(1350) - (70)^2}$$

$$= \frac{843,750 - 806,750}{5400 - 4900} = \frac{37,000}{500}$$

$$= 74$$

Note that there were 4 sets of ordered pairs used in the sample, so $n = 4$. Also, be sure and watch for the difference between $(\Sigma x)^2$ and Σx^2. The denominators for a and b are the same. So once the a term or b term has been calculated some arithmetic can be saved on calculating the other one.

$$b = \frac{4(11,525) - 625(70)}{4(1350) - (70)^2} = \frac{46,100 - 43,750}{500}$$

$$= \frac{2350}{500} = 4.7$$

Inserting these values into the general form, we have

$$y = a + bx = 74 + 4.7x$$

We make the following predictions.

For \$30,000 advertising cost, $x = 30$:

$$y = 74 + 4.7x = 74 + 4.7(30) = 74 + 141 = 215$$

So, an advertising expense of \$30,000 should yield sales of \$215,000.

For \$35,000 advertising cost, $x = 35$:

$$y = 74 + 4.7x = 74 + 4.7(35) = 74 + 164.5 = 238.5$$

So an advertising expense of \$35,000 should yield sales of \$238,500 (see Figure 12.7).

Note that some of the sample data may not be on the line since the line is the best estimate for all the points but need not pass through each sample point. For instance, for the third sample point, the (x, y) values are $(20, 180)$, but from the equation one gets:

$$y = 74 + 4.7x = 74 + 4.7(20) = 74 + 94 = 168, \text{ or } (20, 168)$$

On the other hand, each of the prediction points must fall exactly on the lines because the equation of the line is used to generate the predictions.

This example illustrates that sometimes too many data are available and one must know what is and what is not necessary to work a problem. The best method is to have an orderly approach to the problem and proceed in the same manner on each problem, once you have decided what type problem you are dealing with.

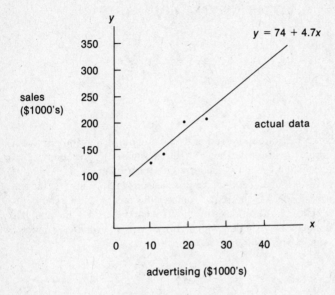

Figure 12.7

13

Analysis of Variance

Variance is a measure of spread or dispersion of the distribution about the mean and is represented as σ^2 (sigma squared). The analysis of variance is a technique that considers the total spread or dispersion of data from several sources, and the results assist us in deciding whether or not there are significant differences among the means of the sources.

In model building, the analysis of variance can be used in the analysis step or to compare a prediction with an occurrence in order to determine if there is significant difference between them.

F TABLE

The differences among means are evaluated by generating an *F statistic*, which is a ratio that compares the variation within the samples to the variation among the samples. This ratio is applied against a standard value of the F statistic that is looked up in the standard table. If the variation within the samples is large compared to the variation among the samples, then the calculated F from the samples will be relatively small compared to the F value from the table (Appendix IV). If the variation among the samples is large compared to the variation within the samples, then the F calculated from the samples will be relatively large compared to the F from the table. How large must one variation be when compared to the other to be significant is handled by the calculations and the table, so one only need know the decision rule:

$$F = \frac{\text{variance among samples}}{\text{variance within samples}}$$

If the calculated F is less than the F from the table, there is not a significant difference between the sample means because the variance among samples is small compared to the variance within each sample. If the calculated F is greater than the F from the table, there is a significant difference because the variance among samples is large compared to the variance within each sample.

FORMULAS

The formula for calculating the variance within a sample is

$$\sigma_x^2 = \frac{\Sigma(x - \bar{x})^2}{\Sigma(N - 1)}$$

where

$\bar{x}$ = mean of the sample

$(x - \bar{x})$ = deviation or difference between a number in the sample and its corresponding mean

$(x - \bar{x})^2$ = deviation squared

$\Sigma(x - \bar{x})^2$ = sum of all the squared deviations of all the samples

N = number of items in the sample, the sample size

$N - 1$ = number of degrees of freedom in the sample, sample size minus 1

$\Sigma(N - 1)$ = sum of all the degrees of freedom of all the samples

The formula for calculating the variance among samples is

$$\sigma_x^2 = \frac{\Sigma(\bar{x} - \bar{\bar{x}})^2}{N - 1}$$

where

$\bar{x}$ = mean of the sample

$\bar{\bar{x}}$ = sum of the sample means divided by the number of samples, population mean

$(\bar{x} - \bar{\bar{x}})$ = deviation or difference among

$(\bar{x} - \bar{\bar{x}})^2$ = deviation squared

$\Sigma(\bar{x} - \bar{\bar{x}})^2$ = sum of all the squared deviations

N = number of samples

$N - 1$ = number of degrees of freedom between the samples, number of samples minus 1

Example 13.1

The following two samples were taken at different times from a process. The question that must be answered is, has the process changed? To answer this, we must determine if there is a significant difference between the average means of the samples.

Sample A	Sample B
20	30
80	40
50	50

(1) First calculate the "within" sample variance.

$$\sigma_x^2 = \frac{\Sigma(x - \bar{x})^2}{\Sigma(N - 1)}$$

We need to know x, $(x - \bar{x})$, $(x - \bar{x})^2$, and $\Sigma(x - \bar{x})^2$ for each sample; the degrees of freedom for each sample; and the total degrees of freedom.

(1a) Find the average means for each of the samples.

Sample A: $\bar{x} = \dfrac{\Sigma x}{N} = \dfrac{20 + 80 + 50}{3} = \dfrac{150}{3} = 50$

Sample B: $\bar{x} = \dfrac{\Sigma x}{N} = \dfrac{30 + 40 + 50}{3} = \dfrac{120}{3} = 40$

(1b) Arrange the data into a table. Compute the deviations from $\bar{x}$ for each sample and then square the deviations.

Sample A, $\bar{x} = 50$

x	$x - \bar{x}$	$(x - \bar{x})^2$
20	-30	900
80	$+30$	900
50	0	0
Total		1800

Sample B, $\bar{x} = 40$

x	$x - \bar{x}$	$(x - \bar{x})^2$
30	-10	100
40	0	0
50	$+10$	100
Total		200

(1c) Calculate the degrees of freedom for each sample, which equals sample size minus 1.

Sample A: $N - 1 = (3 - 1) = 2$
Sample B: $N - 1 = (3 - 1) = 2$

Total degrees of freedom = $\Sigma(N - 1) = 2 + 2 = 4$

(1d) Compute the "within" sample variance.

$$\sigma_x^2 = \frac{\Sigma(x_1 - \bar{x})^2 + \Sigma(x_2 - \bar{x})^2}{(N_1 - 1) + (N_2 - 1)}$$

$$= \frac{1800 + 200}{2 + 2} = \frac{2000}{4}$$

$$= 500$$

(2) Calculate the "among" mean variance.

$$\sigma_{\bar{x}}^2 = \frac{\Sigma(\bar{x} - \bar{\bar{x}})^2}{N - 1}$$

We need to know $\bar{\bar{x}}$ (average of the means of the samples), $\bar{x} - \bar{\bar{x}}$, $(\bar{x} - \bar{\bar{x}})^2$, $\Sigma(\bar{x} - \bar{\bar{x}})^2$, and the number of degrees of freedom.

(2a) Calculate the average of the sample means.

Sample A: $\bar{x} = 50$
Sample B: $\bar{x} = 40$

$$\bar{\bar{x}} = \frac{\Sigma\bar{x}}{N} = \frac{50 + 40}{2} = \frac{90}{2}$$

$$= 45$$

(2b) Arrange the data into a table. Compute the deviations from $\bar{\bar{x}}$ for each sample mean and then square the deviations.

Sample mean	Deviation	Deviation squared
$(\bar{x})$	$(\bar{x} - \bar{\bar{x}})$	$(\bar{x} - \bar{\bar{x}})^2$
50	+5	25
40	−5	25
Total		50

(2c) Calculate the degrees of freedom, which equals the number of samples minus $1 = (2 - 1) = 1$.

(2d) Compute the "among" sample variance.

$$\sigma_{\bar{x}}^2 = \frac{\Sigma(\bar{x} - \bar{\bar{x}})^2}{N - 1}$$

$$\sigma_{\bar{x}}^2 = \frac{50}{1} = 50$$

(3) Compare the "within" sample variance to the "among" sample variance by computing the ratio between them.

$$F = \frac{\text{among sample variance}}{\text{within sample variance}} = \frac{50}{500} = 0.1$$

(4) Compare the computed ratio of 0.1 to the F table value at the "among" sample degrees of freedom $= 1$, and "within" sample degrees of freedom $= 0.4$. (Look up column 1, row 4 in Table 13.1). From the table, the allowable F ratio is $= 7.7$. Since the computed ratio is less than 7.7, there is no significant difference between the two means.

Example 13.2

Grant, of the Consulting Associates, is hired to determine if the 4 plants of Arvey Manufacturing are producing basically the same quality goods. Grant obtains the following samples and then must decide whether the samples are significantly different.

Sample A	Sample B	Sample C	Sample D
8	10	12	5
5	7	9	5
5	6	6	5
	1	5	

(1) Calculate the "within" sample variance.

$$\sigma_{\bar{x}}^2 = \frac{\Sigma(x - \bar{x})^2}{\Sigma(N - 1)}$$

Needed are x, $x - \bar{x}$, $(x - \bar{x})^2$, and $\Sigma(x - \bar{x})^2$ for each sample; the degrees of freedom; and the total degrees of freedom.

(1a) Find the average means for each of the samples.

Sample A: $\bar{x} = \dfrac{\Sigma x}{N} = \dfrac{8 + 5 + 5}{3} = 6$

Sample B: $\bar{x} = \dfrac{\Sigma x}{N} = \dfrac{10 + 7 + 6 + 1}{4} = 6$

Sample C: $\bar{x} = \dfrac{\Sigma x}{N} = \dfrac{12 + 9 + 6 + 5}{4} = 8$

Sample D: $\bar{x} = \dfrac{\Sigma x}{N} = \dfrac{5 + 5 + 5}{3} = 5$

(1b) Calculate the degrees of freedom for each sample. Degrees of freedom equals sample size minus 1 $= N - 1$.

Sample A: $N - 1 = 3 - 1 = 2$
Sample B: $N - 1 = 4 - 1 = 3$

Sample C: $N - 1 = 4 - 1 = 3$
Sample D: $N - 1 = 3 - 1 = 2$

Total degrees of freedom = $\Sigma(N - 1) = 2 + 3 + 3 + 2 = 10$

(1c) Arrange the data into a table, compute the deviations from $\bar{x}$ for each sample, and then square the deviations.

Sample A, $\bar{x} = 6$

x	$x - \bar{x}$	$(x - \bar{x})^2$
8	2	4
5	1	1
5	1	1
Total		6

Sample B, $\bar{x} = 6$

x	$x - \bar{x}$	$(x - \bar{x})^2$
10	4	16
7	1	1
6	0	0
1	−5	25
Total		42

Sample C, $\bar{x} = 8$

x	$x - \bar{x}$	$(x - \bar{x})^2$
12	4	16
9	1	1
6	−2	4
5	−3	9
Total		30

Sample D, $\bar{x} = 5$

x	$x - \bar{x}$	$(x - \bar{x})^2$
5	0	0
5	0	0
5	0	0
Total		0

(1d) Compute the "within" sample variance.

$$\sigma_x^2 = \frac{\Sigma(x_1 - \bar{x})^2 + \Sigma(x_2 - \bar{x})^2 + \Sigma(x_3 - \bar{x})^2 + \Sigma(x_4 - \bar{x})^2}{(N_1 - 1) + (N_2 - 1) + (N_3 - 1) + (N_4 - 1)}$$

$$= \frac{6 + 42 + 30 + 0}{2 + 3 + 3 + 2} = \frac{78}{10} = 7.8$$

(2) Calculate the "among" mean variance.

$$\sigma_{\bar{x}}^2 = \frac{\Sigma(\bar{x} - \bar{\bar{x}})^2}{N - 1}$$

Needed are $\bar{\bar{x}}$ (average of the means of the samples), $\bar{x} - \bar{\bar{x}}$, $(\bar{x} - \bar{\bar{x}})^2$, and the number of degrees of freedom.

(2a) Calculate the average of the means of the samples.

Sample A: $\bar{x}_1 = 6$

Sample B: $\bar{x}_2 = 6$
Sample C: $\bar{x}_3 = 8$
Sample D: $\bar{x}_4 = 5$

$$\bar{\bar{x}} = \frac{\Sigma \bar{x}}{N} = \frac{6 + 6 + 8 + 5}{4} = \frac{25}{4} = 6.25$$

(2b) Arrange the data into a table. Compute the deviations from $\bar{\bar{x}}$ for each sample mean and then square the deviations.

Average of the sample means = $\bar{\bar{x}}$ = 6.25

Sample mean	Deviation	Deviation squared
x	$x - \bar{\bar{x}}$	$(x - \bar{\bar{x}})^2$
6	0.25	0.0625
6	0.25	0.0625
8	1.75	3.0625
5	-1.25	1.5625
Total		4.75

(2c) Calculate the degrees of freedom, which equals the number of samples minus 1 = 4 − 1 = 3.

(2d) Compute the "among" sample variance.

$$\sigma_{\bar{x}}^2 = \frac{\Sigma (\bar{x} - \bar{\bar{x}})^2}{N - 1}$$

$$= \frac{4.75}{3}$$

$$= 1.58$$

(3) Compare the "within" sample variance to the "among" sample variance by computing the ratio between them.

$$F = \frac{\text{"among" sample variance}}{\text{"within" sample variance}} = \frac{1.58}{7.8} = 0.203$$

(4) Compare the computed ratio of 0.203 to the F table value at the "among" sample degrees of freedom = 3, and the "within" sample degrees of freedom = 10. The table ratio is 3.7. Since the computed ratio is *less* than 3.7, there is no significant difference between the means.

Example 13.3

The following results were achieved by 4 people on different tests, which were designated as samples A, B, C, D, and E. The question is whether there are significant differences among the mean scores of the different tests.

Sample A	Sample B	Sample C	Sample D	Sample E
11	22	2	35	46
14	23	3	35	47
15	22	2	35	48
12	21	1	35	43

(1) Calculate the "within" sample variance.

$$\sigma_x^2 = \frac{\Sigma(x - \bar{x})^2}{\Sigma(N - 1)}$$

(1a) Find the average means for each of the samples.

Sample A: $\bar{x} = \dfrac{\Sigma x}{N} = \dfrac{11 + 14 + 15 + 12}{4} = \dfrac{52}{4} = 13$

Sample B: $\bar{x} = \dfrac{\Sigma x}{N} = \dfrac{23 + 23 + 22 + 21}{4} = \dfrac{88}{4} = 22$

Sample C: $\bar{x} = \dfrac{\Sigma x}{N} = \dfrac{2 + 3 + 2 + 1}{4} = \dfrac{8}{4} = 2$

Sample D: $\bar{x} = \dfrac{\Sigma x}{N} = \dfrac{35 + 35 + 35 + 35}{4} = 35$

Sample E: $\bar{x} = \dfrac{\Sigma x}{N} = \dfrac{46 + 47 + 48 + 43}{4} = \dfrac{184}{4} = 46$

(1b) Arrange the data into a table. Compute the deviations from x for each sample and then square the deviations.

Sample A, $\bar{x} = 13$			Sample B, $\bar{x} = 22$		
x	$x - \bar{x}$	$(x - \bar{x})^2$	x	$x - \bar{x}$	$(x - \bar{x})^2$
11	-2	4	22	0	0
14	1	1	23	1	1
15	2	4	22	0	0
12	-1	1	21	-1	1
		10			2

Sample C, $\bar{x} = 2$

x	$x - \bar{x}$	$(x - \bar{x})^2$
2	0	0
3	1	1
2	0	0
1	-1	1
		2

Sample D, $\bar{x} = 35$

x	$x - \bar{x}$	$(x - \bar{x})^2$
35	0	0
35	0	0
35	0	0
35	0	0
		0

Sample E, $\bar{x} = 46$

x	$x - \bar{x}$	$(x - \bar{x})^2$
46	0	0
47	1	1
48	2	4
43	-3	9
		14

(1c) Calculate the degrees of freedom for each sample, which equals sample size minus 1.

Sample A: $N - 1 = 4 - 1 = 3$
Sample B: $N - 1 = 4 - 1 = 3$
Sample C: $N - 1 = 4 - 1 = 3$
Sample D: $N - 1 = 4 - 1 = 3$
Sample E: $N - 1 = 4 - 1 = 3$

Total degrees of freedom = $\Sigma(N - 1) = (3 + 3 + 3 + 3 + 3) = 15$

(1d) Compute the "within" sample variance.

$$\sigma_x^2 = \frac{\Sigma(x - \bar{x})_A^2 + \Sigma(x - \bar{x})_B^2 + \Sigma(x - \bar{x})_C^2 + \Sigma(x - \bar{x})_D^2 + \Sigma(x - \bar{x})_E^2}{(N_A - 1) + (N_B - 1) + (N_C - 1) + (N_D - 1) + (N_E - 1)}$$

$$= \frac{10 + 2 + 2 + 0 + 14}{3 + 3 + 3 + 3 + 3} = \frac{28}{15}$$

$$= 1.87$$

(2) Calculate the "among" mean variance.

$$\sigma_{\bar{x}}^2 = \frac{\Sigma(\bar{x} - \bar{\bar{x}})^2}{N - 1}$$

(2a) Calculate the average of the sample means.

Sample A: $\bar{x} = 13$
Sample B: $\bar{x} = 22$
Sample C: $\bar{x} = 2$
Sample D: $\bar{x} = 35$
Sample E: $\bar{x} = 46$

$$\bar{\bar{x}} = \frac{\Sigma x}{N} = \frac{13 + 22 + 2 + 35 + 46}{5} = \frac{118}{5} = 23.6$$

(2b) Arrange the data into a table, compute the deviations from $\bar{\bar{x}}$ for each sample mean, and then square the deviations.

Average of the sample means, $\bar{\bar{x}} = 23.6$

Sample mean $\bar{x}$	Deviation $\bar{x} - \bar{\bar{x}}$	Deviation squared $(\bar{x} - \bar{\bar{x}})^2$
13	−10.6	112.4
22	−1.6	2.56
2	−21.6	466.6
35	11.4	130
46	22.4	501.8
		1213.4

(2c) Calculate the degrees of freedom, which equals number of samples minus 1 = 5 − 1 = 4.

(2d) Compute the "among" sample variance.

$$\sigma_{\bar{x}}^2 = \frac{1213.4}{4} = 303.4$$

(3) Compare the "within" sample variance to the "among" sample variance by calculating the ratio between them.

$$F = \frac{\text{"among" sample variance}}{\text{"within" sample variance}} = \frac{303.4}{1.87} = 162.2$$

(4) Compare the computed ratio of 162.2 to the F table value at the "among" sample degrees of freedom = 4 and the "within" sample degrees of freedom = 15 (column 4, row .00 in the table). From the table, the allowable F ratio is 2.4. Since the computed ratio is greater than 2.4, there is a significant difference between the means. This is as one might expect upon inspecting the data.

PROBLEM

Problem 13.1

Two methods are employed to predict the sales of Company Z for a certain period, and then the actual records for the period are produced. You wish to know if there is a significant difference between the three averages; i.e., those from the two methods of predicting and the actual figures (all figures in $1000's).

Prediction A	Prediction B	Actual figures (Sample C)
25	10	15
30	15	20
35	20	25

(1) Calculate the "within" sample variance.

$$\sigma_x^2 = \frac{\Sigma(x - \bar{x})^2}{\Sigma(N - 1)}$$

We need to find x, $(x - \bar{x})$, $(x - \bar{x})^2$, and $\Sigma(x - \bar{x})^2$ for each sample.

(1a) Find the average means for each sample.

Sample A: $\bar{x} = \dfrac{\Sigma x}{N} = \dfrac{25 + 30 + 35}{3} = 30$

Sample B: $\bar{x} = \dfrac{\Sigma x}{N} = \dfrac{10 + 15 + 20}{3} = 15$

Sample C: $\bar{x} = \dfrac{\Sigma x}{N} = \dfrac{15 + 20 + 25}{3} = 20$

(1b) Arrange the data into a table and compute deviations.

Sample A

x	$x - \bar{x}$	$(x - \bar{x})^2$
25	-5	25
30	0	0
35	5	25
		50

Sample B

x	$x - \bar{x}$	$(x - \bar{x})^2$
10	−5	25
15	0	0
20	5	25
		50

Sample C

x	$x - \bar{x}$	$(x - \bar{x})^2$
15	−5	25
20	0	0
25	5	25
		50

(1c) Calculate the degrees of freedom for each sample which equals sample size minus 1.

Sample A: $N - 1 = 3 - 1 = 2$
Sample B: $N - 1 = 3 - 1 = 2$
Sample C: $N - 1 = 3 - 1 = 2$

Total degrees of freedom $= \Sigma (N - 1) = 2 + 2 + 2 = 6$

(1d) Compute the "within" sample variance.

$$\frac{(x - \bar{x})^2}{6} = \frac{50 + 50 + 50}{6} = \frac{150}{6}$$

$$= 25$$

(2) Calculate the "among" mean variance.

$$\sigma_{\bar{x}}^2 = \frac{\Sigma(\bar{x} = \bar{\bar{x}})^2}{N - 1}$$

(2a) Calculate the average of the sample means.

Sample A: $\bar{x} = 30$
Sample B: $\bar{x} = 15$
Sample C: $\bar{x} = 20$

$$\bar{x} = \frac{\Sigma \bar{x}}{N} = \frac{30 + 15 + 20}{3} = \frac{65}{3} = 21.7$$

(2b) Arrange the data into a table. Compute the deviations from $\bar{x}$ for each sample mean and square the deviations.

Sample mean	Deviation	Deviation squared
30	8.3	68.9
15	−6.7	44.9
20	−1.7	2.9
		116.7

(2c) Calculate the degrees of freedom, which equals
$$N - 1 = 3 - 1 = 2$$

(2d) Compute the "among" sample variance.

$$\sigma_{\bar{x}}^2 = \frac{116.7}{2} = 58.4$$

(3) Compare the within sample variance to the between sample variance by computing the ratio between them.

$$F = \frac{\text{"among" sample variance}}{\text{"within" sample variance}} = \frac{58.4}{25} = 2.34$$

(4) Compare the ratio of 2.34 to the F table value at the "among" sample degrees of freedom = 2, and the "within" sample degrees of freedom = 6. From the table, the allowable F ratio is 5.1. Since the computed ratio is less than 5.1, there is no significant difference between the means.

14

Continuous Probability
Distributions

There are some experiments for which it is impossible to list all the possible outcomes in the sample space before they are performed, because there are an infinite number of possible outcomes. For instance, the measurement of someone's height depends on at least the following: who is doing the measuring, the accuracy of the measuring device, the gradation of the measuring device, the time of the day of the measuring, whether one is using the metric system, and human error. In discrete-style problems each possible outcome can be listed in the sample space, and often the probability of each particular outcome can be generated.

Continuous random variables are used when the sample space cannot be listed outcome by outcome and when the range in which the outcomes might lie can be defined. In addition, the probability of any particular outcome is always 0. The probability that an outcome may lie within a specified range may be known or estimated. For example, in continuous problems the probability that $x = 2$ is always 0, but, depending on the distribution, the probability that x is between 1 and 3 may not be 0.

Continuous distributions may be used effectively when many outcomes may be achieved that are close in numerical value to each other but that rarely or ever are exactly the same value. Examples of discrete and continuous functions are graphed in Figure 14.1.

In continuous random variable problems the range in which the variable may assume values other than 0 is defined and the probability that the variable may be within some bounds may be ascertained.

A continuous problem can be identified by the characteristic that the range in which the variable does not equal 0 must be listed, for example, $0 < x < 5$ or $1 \leq w \leq 6$. The value of the function for any region outside the area in which the function is specifically listed not to be 0 is by definition 0. For example, if the function is defined as not 0 for $0 \leq x \leq 5$, then for $x > 5$ or $x < 0$ the function is 0.

To determine the probability of a variable being between two points, the area under the curve between the points must be determined. The

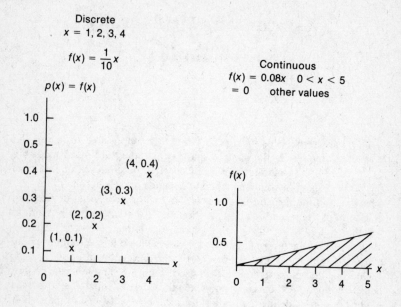

Figure 14.1

area under the curve for the region in which the function is defined not to be 0 must be 1 in order for the function to be a probability function. In addition, the area between any two points is a percentage of 1 and represents the probability that the variable is between the points.

The area under a curve is obtained by using a technique taught in calculus called "integrating" a function between the points concerned with.

GENERAL FORM OF INTEGRATION

$$\int_c^d ax^b = a \left. \frac{x^{b+1}}{b+1} \right|_c^d = a \left(\frac{1}{b+1} \right)(d^{b+1} - c^{b+1})$$

where a, b, c, d are constants that will be defined as numbers for each problem, x^b is a variable, and $\int$ indicates integration. The letter at the bottom (c) indicates the starting point and the letter at the top (d) indicates the other end of the bounds over which the function is being integrated. The coefficient (a) is rewritten and the exponent (b) has 1 added to it and the term is then divided by the new exponent. The new expression must be evaluated from the lower bound to the upper bound by first inserting the upper bound for the variable and getting a numerical evaluation, then repeating this for the lower bound and subtracting this numerical evaluation from the one generated by the upper bound. Inserting the upper bound gives the area under the curve from 0 to the lower bound, and subtracting the latter from the former leaves the area between the bounds, or the desired result.

For example, we insert the following into the general form of the integration formula and graph the function in Figure 14.2.

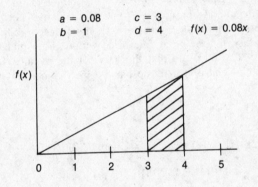

Figure 14.2

$$A(0 < x < 3) = \int_0^3 0.08x$$

$$A(0 < x < 4) = \int_0^4 0.08x$$

$$P(3 < x < 4) = \int_3^4 0.08x$$

$$= \int_0^4 0.08x - \int_0^3 0.08x$$

$$= 0.08 \left.\frac{x^2}{2}\right|_0^4 - 0.08 \left.\frac{x^2}{2}\right|_0^3$$

or, putting it into one step,

$$P(3 < x < 4) = \int_3^4 0.08x^1 = 0.08 \left. \frac{x^2}{2} \right|_3^4 = 0.04x^2 \left. \right|_3^4$$

$$= 0.04(4)^2 - 0.04(3)^2 = 0.04(16 - 9) = 0.28$$

Thus, 0.28 is the probability of the variable value being between 3 and 4.

Because the area under the curve at exactly some point is 0, we have $P(a < x < b) = P(a \leq x \leq b)$.

The only difference between the two expressions is that the one on the left does not include the endpoints (a and b) and the one on the right does include them. However, because the probability of exactly any point is 0, the endpoints don't contribute anything toward the probability; the numerical calculation is the same whether or not they are included. This is true only for continuous functions and not for discrete functions.

To calculate *the cumulative probability function* at any point, the function is integrated from the lowest value for which it is defined not to be 0 to the point concerned with. The area under the curve between the lowest boundary and the point concerned with is the cumulative at the point concerned with. Note that the value of the cumulative must be between 0 and 1 and is usually designated by a capital letter.

Example 14.1

$$F(2) = \int_0^2 0.08x = 0.08 \left. \frac{x^2}{2} \right|_0^2 = 0.04(2)^2 - 0.04(0)^2$$

$$= 0.04(4) - 0 = 0.16$$

To determine the cumulative for any particular point, the integration would have to be repeated for that point. For some problems the generation of the cumulative for many points may be simplified by first solving for the cumulative in terms of x and then substituting in the desired points, which entails only one necessary integration.

Example 14.2

$$F(x) = \int_0^x 0.08x = 0.08 \left. \frac{x^2}{2} \right|_0^x = 0.04(x)^2 - 0.04(0)^2 = 0.04x^2$$

Now, to find the cumulative for any point, merely insert the values for the point. Generating the cumulative in terms of x also enables another style of problem to be approached. Suppose one knows the value of the cumulative at some point, the percent that scored below the point, and wishes to know the point at which this occurs, in other words, a reverse of the original style problem. Where in the original problem the values of x were given and the cumulative for the points were determined, in the new approach the cumulative is given and the value of x at which it must occur is sought. To solve this type of problem generate the cumulative in terms of x and set this equal to the value of the cumulative concerned with. Then solve this equation for x.

Example 14.3

Based on the previous examples, you are told that a score is the 80th percentile (80% score below it). What is the score?

$$0.80 = 0.04x^2$$
$$x = 4.47$$

Thus, a score of 4.47 will have 80% of the scores below it.

EXPECTED VALUE, VARIANCE, AND STANDARD DEVIATION

The function $f(x)$ can be used in formulas for the expected value and variance, with integration being performed instead of summation. For the *expected value*, the formula becomes

$$E(x) = \int_{\text{lower bound}}^{\text{upper bound}} x f(x) dx$$

Note that the evaluation technique is the same as illustrated in the previous probability problems. Consider the previous example:

$$E(x) = \int_0^5 x f(x) = \int_0^5 x(0.08x) = \int_0^5 0.08x^2$$

$$= \frac{0.08x^3}{3} \Big|_0^5 = \frac{0.08}{3}(5)^3 - 0 = 3.33$$

The expected value is interpreted in the same way as in discrete probability problems.

The *variance* also maintains the same basic formula for discrete sample spaces, with the only alteration again being the substitution of the operation of integration for the summation operation.

$$\text{variance} = \int_{\text{lower bound}}^{\text{upper bound}} [x - E(x)]^2 f(x)$$

Again considering the same example, we have

$$\text{variance} = \int_0^5 (x - 3.33)^2 \, 0.08x = \int_0^5 [x^2 - 6.66x + (3.33)^2] \, 0.08x$$

$$= \int_0^5 0.08x^3 - 53x^2 + 0.89x$$

$$= \frac{0.08x^4}{4} - \frac{0.53x^3}{3} + \frac{0.89x^2}{2} \Big|_0^5 = 0.02(5)^4 - \frac{0.53}{3}(5)^3$$

$$+ \frac{0.89(5)^2}{2} - 0$$

$$= 12.5 - 22.1 + 11.1 - 0 = 1.5$$

$$\text{Standard deviation} = \sqrt{\text{variance}} = \sqrt{1.5} = 1.2$$

Be sure to square the terms in the variance formula, where indicated, before performing the integration. Also be sure to subtract the lower bound, which in this case was 0, although it need not always be 0.

Example 14.4

$$f(x) = 3x^2 \qquad 0 \le x \le 1$$
$$= 0 \qquad \text{elsewhere}$$

This function represents customers' arrival at a check-out counter per minute. What is the probability of an arrival during the various times indicated?

(a) In the first 15 sec (0.25 min) of a minute

$$P(0 < x < 0.25) = \int_0^{0.25} 3x^2 = \frac{3x^3}{3} \Big|_0^{0.25} (0.25)^3 - (0)^3 = 0.02$$

(b) In the last 15 sec of a minute

$$P(0.75 < x < 1) = \int_{0.75}^1 3x^2 = \frac{3x^3}{3} \Big|_{0.75}^1 = (1)^3 - (0.75)^3 = 0.58$$

(c) Between 15 and 45 sec of a minute

$$P(0.25 < x < 0.75) = \int_{0.25}^{0.75} 3x^2 = \frac{3x^3}{3} \Big|_{0.25}^{0.75} = (0.75)^3 - (0.25)^3 = 0.40$$

Note that when the time spans covered in Parts (a), (b), (c) are combined, they cover all possible times; and when their probabilities are added, they total 1, as they must in order for this to be a probability function.

(d) What is the general form of the cumulative function?

$$F(x) = \int_0^x 3x^2 = \frac{3x^3}{3} \Big|_0^x = x^3 - 0 = x^3$$

(e) At what time (x) will 60% of the customers have arrived? (Set the cumulative equal to 0.60 and solve.)

$$F(x) = x^3 = 0.60$$

$$x = 0.84$$

That is, at 0.84 min, or 50.4 sec, 60% of the customers will have arrived.

(f)   What is the expected number of arrivals per minute?

$$E(x) = \int_0^1 xf(x) = \int_0^1 x \, 3x^2 = \int_0^1 3x^3 = \frac{3x^4}{4} \Big|_0^1 = 0.75 - 0 = 0.75$$

(g)   What is the variance?

$$\text{variance} = \int_0^1 [x - E(x)]^2 f(x) = \int_0^1 (x - 0.75)^2 \, 3x^2$$

$$= \int_0^1 (x^2 - 1.5x + 0.56) \, 3x^2 = \int_0^1 3x^4 - 4.5x^3 + 1.7x^2$$

$$= \frac{3}{5} x^5 - \frac{4.5}{4} x^4 + \frac{1.7}{3} x^3 \Big|_0^1 = 0.6 - 1.13 + 0.57$$

$$= 0.04$$

(h)   What is the standard deviation?

$$\text{standard deviation} = \sqrt{\text{variance}} = \sqrt{0.04} = 0.2$$

Example 14.5

The function

$$f(x) = \frac{1}{2} - \frac{1}{8} x \qquad 0 \le x \le 4$$

$$= 0 \qquad\qquad \text{elsewhere}$$

represents the probability of defects in 30 yd^2 of cloth.

(a)   What is the expected number of defects?

$$E(x) = \int_0^4 xf(x) = \int_0^4 x \left(\frac{1}{2} - \frac{1}{8} x \right) = \int_0^4 \frac{1}{2} x - \frac{1}{8} x^2$$

$$= \left(\frac{1}{2} \right) \frac{x^2}{2} - \left(\frac{1}{8} \right) \frac{x^3}{3} \Big|_0^4 = \frac{4^2}{4} - \frac{4^3}{24} - 0 = 1.33$$

(b)   What is the variance?

$$\text{variance} = \int_0^4 [x - E(x)]^2 f(x) = \int_0^4 (x - 1.33)^2 \left(\frac{1}{2} - \frac{1}{8} x \right)$$

$$= \int_0^4 (x^2 - 2.66x + 1.77)(0.5 - 0.125x)$$

$$= \int_0^4 0.125x^3 + 0.83x^2 - 1.55x + 0.885$$

$$= \frac{0.125(4)^3}{4} + \frac{0.83(4)^3}{3} - \frac{1.55(4)^2}{2} + 0.885(4) - 0$$

$$= 2 + 17.7 - 12.4 + 3.54$$

$$= 10.84$$

(c) What is the standard deviation?

$$\text{standard deviation} = \sqrt{\text{variance}} = \sqrt{10.84} = 3.29$$

(d) What is the probability of between 1 and 2 defects?

$$P(1 < x < 2) = \int_1^2 \frac{1}{2} - \frac{1}{8}x = \frac{1}{2}x - \left(\frac{1}{8}\right)\frac{x^2}{2} \Big|_1^2$$

$$= \frac{1}{2}(2) - \left(\frac{1}{8}\right)\frac{(2)^2}{2} - \left(\frac{1}{2}(1) - \frac{1}{8}\frac{(1)^2}{2}\right) = \frac{5}{16}$$

$$= 0.31$$

PROBLEMS

Problem 14.1

Given a function $f(x) = 0.5x$ for $1 \leq x \leq \sqrt{5}$ and $f(x) = 0$ elsewhere, determine the following:

(a) Is it a probability function?

To be a probability function, the following must be true:

$$\int_{\text{lower bound}}^{\text{upper bound}} f(x) = 1$$

Since $1 \leq x \leq \sqrt{5}$, for the function to be a probability function, the following must be true:

$$\int_1^{\sqrt{5}} 0.5x = 1$$

Is it true?

$$\int_1^{\sqrt{5}} 0.5x = \frac{0.5x^2}{2} \Big|_1^{\sqrt{5}} = \frac{0.5(\sqrt{5})^2}{2} - \frac{0.5(1)^2}{2}$$

$$= \frac{0.5(5)}{2} - \frac{0.5(1)}{2} = \frac{2.5 - 0.5}{2}$$

$$= 1$$

So the function is a probability function.

(b) What is the probability of x being less than 1.5?

$$\int_1^{1.5} 0.5x = \frac{0.5x^2}{2} \Big|_1^{1.5} = \frac{0.5(1.5)^2}{2} - \frac{0.5(1)^2}{2}$$

$$= \frac{0.5(2.25) - 0.5(1)}{2} = \frac{0.5(1.25)}{2}$$

$$= 0.3125$$

Note that the function was integrated from the lower bound, which in this case was 1, and not from 0.

(c) What is the probability of x being greater than 1.5?

There are two ways to do this and both should give the same results. Both are shown here.

(1)

$$P(x < 1.5) = 1 - \int_{\text{lower bound}}^{\text{point asked about}} f(x)$$

$$= 1 - \int_1^{1.5} 0.5x = 1 - \frac{0.5x^2}{2} \Big|_1^{1.5}$$

$$= 1 - \left(\frac{0.5(1.5)^2}{2} - \frac{0.5(1)^2}{2} \right) = 1 - 0.3125$$

$$= 0.6875$$

(2)

$$P(x < 1.5) = \int_{\text{point concerned with}}^{\text{upper bound}} f(x)$$

$$= \int_{1.5}^{\sqrt{5}} 0.5x = \frac{0.5x^2}{2} \Big|_{1.5}^{\sqrt{5}}$$

$$= \frac{0.5(\sqrt{5})^2}{2} - \frac{0.5(1.5)^2}{2} = \frac{0.5(5 - 2.25)}{2}$$

$$= 0.6875$$

Note that both methods yield the same result.

(d) What is the probability of $1.3 < x < 2$?

$$P(1.3 < x < 2) = \int_{1.3}^2 0.5x = \frac{0.5x^2}{2} \Big|_{1.3}^2$$

$$= \frac{0.5(2)^2}{2} - \frac{0.5(1.3)^2}{2}$$

$$= \frac{0.5(4)}{2} - \frac{0.5(1.69)}{2}$$

$$= 1 - 0.4225$$

$$= 0.5775$$

(e) Derive the cumulative function in terms of x.

$$F(x) = \int_1^x 0.5x = \frac{0.5x^2}{2} \Big|_1^x = \frac{0.5x^2}{2} - \frac{0.5(1)}{2} = 0.25x^2 - 0.25$$

(f) What is the value of the cumulative function at $x = 2$?

Two methods can be used.

(1)

Use the result from Part (e).

$$F(x = 2) = 0.25(2)^2 - 0.25 = 1 - 0.25 = 0.75$$

This is the easier method if the cumulative function in terms of x has already been generated.

(2)

$$F(x = 2) = \int_1^2 0.5x = \frac{0.5x^2}{2} \Big|_1^2 = \frac{0.5(2)^2}{2} - \frac{0.5(1)^2}{2} = 0.75$$

Note both answers are identical.

(g) What is the expected value of x?

$$E(x) = \int_1^{\sqrt{5}} 0.5x(x) = \int_1^{\sqrt{5}} 0.5x^2 = \frac{0.5x^3}{3} \Big|_1^{\sqrt{5}} = \frac{0.5(\sqrt{5})^3}{3} - \frac{0.5(1)^3}{3}$$

$$\doteq \frac{0.5(11.18)}{3} - \frac{0.5(1)}{3} = 1.88 - 0.17$$

$$= 1.69$$

(h) What is the variance of x?

$$\text{variance} = \int_1^{\sqrt{5}} [x - E(x)]^2 f(x) = \int_1^{\sqrt{5}} (x - 1.7)^2 \, 0.5x$$

$$= \int_1^{\sqrt{5}} (x^2 - 3.4x + 2.89) \, 05x = \int_1^{\sqrt{5}} 0.5x^3 - 1.7x^2 + 1.445x$$

$$= \frac{0.5x^4}{4} - \frac{1.7x^3}{3} + \frac{1.445x^2}{2} \Big|_1^{\sqrt{5}}$$

$$= \frac{0.5(\sqrt{5})^4}{4} - \frac{1.7(\sqrt{5})^3}{3} + \frac{1.445(\sqrt{5})^2}{2}$$

$$-\left(\frac{0.5(1)^4}{4} - \frac{1.7(1)^3}{3} + \frac{1.445(1)^2}{2}\right)$$

$$= \frac{0.5(25)}{4} - \frac{1.7(11.2)}{3} + \frac{1.445(5)}{2} - \left(\frac{5}{4} - \frac{1.7}{3} + \frac{1.445}{2}\right)$$

$$= 3.125 - 6.346 + 3.613 - (0.125 - 0.567 + 0.723)$$

$$= 0.111$$

(i) What is the standard deviation?

$$\text{standard deviation} = \sqrt{\text{variance}}$$

$$= \sqrt{0.111} = 0.333$$

Problem 14.2

Given a function $f(x) = 0.1x$ for $2 < x < \sqrt{24}$ and $f(x) = 0$ elsewhere, determine the following:

(a) Is it a probability function?

To be a probability function, the relation must be true

$$\int_{\text{lower bound}}^{\text{upper bound}} f(x) = 1$$

For this function to be a probability function, it must be that

$$\int_{2}^{\sqrt{24}} 0.1x = 1$$

The question is, does this happen?

$$\int_{2}^{\sqrt{24}} 0.1x = \frac{0.1x^2}{2} \bigg|_{2}^{\sqrt{24}} = \frac{0.1(\sqrt{24})^2}{2} - \frac{0.1(2)^2}{2} = 1$$

The function is a probability function.

(b) What is the probability of x being less than 3?

$$P(x < 3) = \int_{2}^{3} 0.1x = \frac{0.1x^2}{2} \bigg|_{2}^{3} \frac{0.1(3)^2}{2} - \frac{0.1(2)^2}{2}$$

$$= 0.45 - 0.2 = 0.25$$

(c) What is the probability of x being between 2.5 and 3.5?

$$P(2.5 < x < 3.5) = \int_{2.5}^{3.5} 0.1x = \frac{0.1x^2}{2} \bigg|_{2.5}^{3.5} = \frac{0.1(3.5)^2}{2} - \frac{0.1(2.5)^2}{2}$$

$$= 0.30$$

(d)   What is the expected value of x?

$$E(x) = \int_2^{\sqrt{24}} x(0.1x) = \int_2^{\sqrt{24}} 0.1x^2 = \frac{0.1x^3}{3} \bigg|_2^{\sqrt{24}}$$

$$= \frac{0.1(\sqrt{24})^3}{3} - \frac{0.1(2)^3}{3} = 3.92 - 0.27$$

$$= 3.65$$

(e)   What is the variance?

$$\text{variance} = \int_2^{\sqrt{24}} [x - E(x)]^2\, f(x) = \int_2^{\sqrt{24}} (x - 3.65)^2\, 0.1x$$

$$= \int_2^{\sqrt{24}} (x^2 - 7.3x + 13.32)\, 0.1x = \int_2^{\sqrt{24}} 0.1x^2 - 0.73x + 1.332$$

$$= \frac{0.1x^3}{3} - \frac{0.73x^2}{2} + 1.332x \bigg|_2^{\sqrt{24}}$$

$$= \frac{0.1(\sqrt{24})^3}{3} - \frac{0.73(\sqrt{24})^2}{2} + 1.332(\sqrt{24})$$

$$- \left(\frac{0.1(2)^3}{3} - \frac{0.73(2)^2}{2} + 1.332(2) \right)$$

$$= 1.676$$

(f)   What is the standard deviation?

$$\text{standard deviation} = \sqrt{\text{variance}} = \sqrt{1.676} = 1.29$$

Problem 14.3

Given a function $f(x) = 0.2x$ for $0 < x < 4$ and $f(x) = 0$ elsewhere, is it a probability function?

To be a probability function it must be true that

$$\int_{\text{lower bound}}^{\text{upper bound}} f(x) = 1$$

For this problem for the function to be a probability function, we must have

$$\int_0^4 0.2x = 1$$

The question is, does this happen?

$$\int_0^4 0.2x = \frac{0.2x^2}{2} \bigg|_0^4 = \frac{0.2(4)^2}{2} - \frac{0.2(0)^2}{2} = 6.4 - 0 \neq 1$$

The function integrated over the given bounds does not equal 1, so the function is not a probability function.

APPENDIX I
z TABLE—AREAS UNDER THE NORMAL CURVE

$z = \dfrac{x - \mu}{\sigma}$	Second decimal place in z									
	.00	.01	.02	.03	.04	.05	.06	.07	.08	.09
0.0	.0000	.0040	.0080	.0120	.0160	.0199	.0239	.0279	.0319	.0359
0.1	.0398	.0438	.0478	.0517	.0557	.0596	.0636	.0675	.0714	.0753
0.2	.0793	.0832	.0872	.0910	.0948	.0987	.1026	.1064	.1103	.1141
0.3	.1179	.1217	.1255	.1293	.1332	.1368	.1406	.1443	.1480	.1518
0.4	.1554	.1591	.1628	.1664	.1700	.1736	.1772	.1808	.1844	.1879
0.5	.1915	.1950	.1985	.2019	.2054	.2088	.2123	.2157	.2190	.2224
0.6	.2257	.2291	.2324	.2357	.2389	.2422	.2454	.2486	.2517	.2549
0.7	.2580	.2611	.2642	.2672	.2704	.2734	.2764	.2793	.2823	.2852
0.8	.2882	.2910	.2939	.2967	.2995	.3023	.3051	.3078	.3106	.3133
0.9	.3159	.3185	.3212	.3238	.3264	.3289	.3315	.3340	.3365	.3389
1.0	.3413	.3438	.3462	.3485	.3508	.3531	.3554	.3577	.3599	.3621
1.1	.3643	.3665	.3686	.3708	.3729	.3749	.3770	.3790	.3810	.3830
1.2	.3849	.3869	.3888	.3907	.3925	.3943	.3962	.3980	.3997	.4015
1.3	.4033	.4049	.4066	.4082	.4099	.4115	.4131	.4147	.4162	.4177
1.4	.4192	.4207	.4222	.4236	.4250	.4265	.4279	.4292	.4306	.4319
1.5	.4332	.4345	.4357	.4370	.4382	.4394	.4406	.4418	.4429	.4441
1.6	.4452	.4462	.4474	.4484	.4495	.4505	.4515	.4525	.4535	.4545
1.7	.4554	.4564	.4573	.4582	.4591	.4599	.4608	.4616	.4625	.4633
1.8	.4641	.4649	.4656	.4663	.4671	.4678	.4686	.4693	.4699	.4706
1.9	.4712	.4719	.4726	.4732	.4738	.4744	.4750	.4756	.4760	.4767
2.0	.4772	.4778	.4783	.4788	.4793	.4798	.4803	.4808	.4812	.4818
2.1	.4821	.4826	.4830	.4834	.4838	.4842	.4846	.4850	.4854	.4857
2.2	.4861	.4865	.4868	.4871	.4875	.4878	.4882	.4884	.4887	.4890
2.3	.4893	.4896	.4898	.4901	.4903	.4906	.4909	.4911	.4913	.4916
2.4	.4918	.4920	.4922	.4925	.4927	.4929	.4931	.4932	.4934	.4936
2.5	.4938	.4940	.4941	.4943	.4945	.4946	.4948	.4949	.4951	.4952
2.6	.4953	.4955	.4956	.4958	.4959	.4960	.4961	.4961	.4963	.4964
2.7	.4965	.4966	.4967	.4968	.4969	.4970	.4971	.4972	.4973	.4974
2.8	.4974	.4975	.4976	.4977	.4977	.4978	.4979	.4979	.4980	.4980
2.9	.4980	.4982	.4982	.4983	.4984	.4984	.4985	.4985	.4986	.4986
3.0	.4987	.4987	.4988	.4988	.4988	.4989	.4989	.4990	.4990	.4990
3.9	.5000	.5000	.5000	.5000	.5000	.5000	.5000	.5000	.5000	.5000

First decimal place in z

APPENDIX II
t DISTRIBUTION TABLE

Degrees of freedom	Confidence Levels		
	90%	95%	99%
1	6.31	12.71	63.66
2	2.92	4.30	9.93
3	2.35	3.18	5.84
4	2.13	2.78	4.60
5	2.02	2.57	4.03
6	1.94	2.45	3.71
7	1.89	2.37	3.50
8	1.86	2.31	3.36
9	1.83	2.26	3.25
10	1.81	2.23	3.17
11	1.79	2.20	3.10
12	1.78	2.18	3.06
13	1.77	2.16	3.01
14	1.76	2.15	2.98
15	1.75	2.13	2.95
16	1.74	2.12	2.92
17	1.74	2.11	2.90
18	1.73	2.10	2.88
19	1.73	2.09	2.86
20	1.72	2.09	2.84
21	1.72	2.08	2.83
22	1.72	2.07	2.82
23	1.71	2.07	2.81
24	1.71	2.06	2.80
25	1.71	2.06	2.79
	1.65	1.96	2.56

APPENDIX III
χ^2 STATISTIC

Degrees of freedom	Selected confidence levels		
	0.99	0.95	0.05
1	6.64	3.85	0.004
2	9.21	5.99	0.10
3	11.35	7.82	0.35
4	13.28	9.49	0.70
5	15.09	11.07	1.15
6	16.81	12.59	1.64
7	18.48	14.07	2.17
8	20.09	15.51	2.73
9	21.67	16.92	3.33
10	23.21	18.31	3.94

APPENDIX IV
F TABLE

F		Degrees of freedom—variance among samples										
		1	2	3	4	5	6	7	8	9	10	∞
grees of	1	161	200	216	225	230	234	237	239	241	242	254
edom—	2	18.5	19.0	19.2	19.2	19.3	19.3	19.4	19.4	19.4	19.4	19.5
riance	3	10.2	9.5	9.3	9.1	9.0	8.9	8.9	8.9	8.8	8.8	8.5
:hin	4	7.7	6.9	6.6	6.4	6.3	6.2	6.1	6.0	6.0	6.0	5.6
nples	5	6.6	5.8	5.4	5.2	5.1	5.0	4.9	4.8	4.8	4.7	4.4
	6	6.0	5.1	4.8	4.5	4.4	4.3	4.2	4.2	4.1	4.1	3.7
	7	5.6	4.7	4.4	4.1	4.0	3.9	3.8	3.7	3.7	3.6	3.2
	8	5.3	4.5	4.1	3.8	3.7	3.6	3.5	3.4	3.4	3.4	2.9
	9	5.1	4.3	3.9	3.6	3.5	3.4	3.3	3.2	3.2	3.1	2.7
	10	5.0	4.1	3.7	3.5	3.3	3.2	3.1	3.1	3.0	3.0	2.5
	∞	3.8	3.0	2.6	2.4	2.2	2.1	2.0	1.9	1.9	1.8	1.0

Index